A Practical Manual of Pharmaceutical Engineering

A Practical Manual of Pharmaceutical Engineering

By

P.S. SONA

M Pharm

Associate Professor, Department of Pharmaceutics
Parul Institute of Pharmacy, Limda, Waghodia
Vadodara, Gujarat

UNIVERSITY SCIENCE PRESS

(An Imprint of Laxmi Publications Pvt. Ltd.)

An ISO 9001:2008 Company

BENGALURU • CHENNAI • COCHIN • GUWAHATI • HYDERABAD
JALANDHAR • KOLKATA • LUCKNOW • MUMBAI • RANCHI • NEW DELHI
BOSTON (USA) • ACCRA (GHANA) • NAIROBI (KENYA)

A PRACTICAL MANUAL OF PHARMACEUTICAL ENGINEERING

Printed and bound in India
Typeset at Shubham Composer, Delhi
First Edition : 2015
ISBN 978-93-83828-43-2

Branches

✆	Bengaluru	080-26 75 69 30	
✆	Chennai	044-24 34 47 26,	24 35 95 07
✆	Cochin	0484-237 70 04,	405 13 03
✆	Guwahati	0361-254 36 69,	251 38 81
✆	Hyderabad	040-27 55 53 83,	27 55 53 93
✆	Jalandhar	0181-222 12 72	
✆	Kolkata	033-22 27 43 84	
✆	Lucknow	0522-220 99 16	
✆	Mumbai	022-24 91 54 15,	24 92 78 69
✆	Ranchi	0651-220 44 64	

PUBLISHED IN INDIA BY

UNIVERSITY SCIENCE PRESS
(An Imprint of Laxmi Publications Pvt.Ltd.)
An ISO 9001:2008 Company
113, GOLDEN HOUSE, DARYAGANJ,
NEW DELHI - 110002, INDIA
Telephone : 91-11-4353 2500, 4353 2501
Fax : 91-11-2325 2572, 4353 2528
www.laxmipublications.com info@laxmipublications.com

C— 9941/015/03
Printed at: Ajit Printing Press - Delhi

Dedication

Dedicated to my parents

and

my daughter

PREFACE

This book has been written with an intention to cover all the possible experiments which are to be conducted in the pharmaceutical engineering laboratory at the undergraduate level. I have tried to incorporate all the experiments suggested under pharmaceutical engineering by different universities. Though I have taken long time to complete this book, I have no regrets because the designed experiments are all practically performed in the laboratory by my students and that has given me ample chance to improve the quality of the experiments. During this period I could observe the difficulties of the students in collecting primary information which is the part of the main experiments. The usage of vernier calipers, screw gauge, colorimetry, different standard value like specific heat and radiation constants of different materials, and conversion of units are examples. I have included all such information in the appendices so that students are benefited to get them in a single book.

Today I am very proud to present before you my book "*A Practical Manual of Pharmaceutical Engineering*". Hope that it will be well accepted by the pharmaceutical science fraternity.

The suggestions for the improvement of the book in the forthcoming edition will be gratefully acknowledged.

—Author

ACKNOWLEDGEMENT

I express my indebtedness to my husband Mr. C. Muthulingam who has been an inspiration and support for completing the book.

I extend my sincere thanks to my students of Grace College of Pharmacy (First Batch) and Ramgopal College of Pharmacy (First and Second Batch) for their support.

I thank my brother Mr. Sujith. P.S, B Tech, for his valuable suggestions.

I express my gratitude to the publisher for publishing my book.

—Author

CONTENTS

EXPERIMENT 1

DETERMINATION OF RADIATION CONSTANT OF METAL CYLINDER

AIM

To determine the radiation constant of the given metal cylinder.

To demonstrate the cooling pattern during natural cooling of a hot body.

REQUIREMENTS

Metal cylinder (1)

(made up of any metal for example iron, aluminum, brass, copper of 1 inch diameter, 1.5 inch length which is having drilled hole half way down, so that thermometer would fit nicely).

Hot plate (1)

Digital thermometer (1)

Vernier calipers (1)

Weighing balance (1)

Weight box (1)

Tongs (1)

Hot pad (1)

PRINCIPLE

Heat transfer process plays an important role in science and engineering. When two objects at different temperature are brought into contact, heat flow occurs from the object having higher temperature to the object having lower temperature. The total heat loss from the body to its surroundings often includes appreciable losses by conduction, convection, and radiation. If the body is not in contact with any surfaces the heat transfer by conduction is neglected. So the heat transfer is mainly by convection and radiation. The present experiment demonstrates the pattern of cooling of a hot body by combined convection and radiation.

Radiative cooling is described by Stefan-Boltzmann Law which states that the rate of heat loss per unit surface area of a body at temperature T is proportional to $T^4 - T_s^4$ (T_s is temperature of the surrounding). The rate of heat loss due to combined convection and radiation can be determined by the equation,

$$M\frac{dT}{dt} = \frac{h_1 S}{C}(T - T_{air}) + \frac{h_2 S}{C}(T^4 - T_s^4) \quad \text{...(Eq. 1.1)}$$

where,

M = Mass of the body (Kg)

dT/dt = The rate of change of temperature (Rate of heat loss) (K/sec)

h_1 = Heat transfer coefficient for pure convection. (W/ m^2.K)

S = Surface area of the hot body (m^2)

T = Temperature of the metal surface (K)

T_{air} = Temperature of the surroundings (K)

T_s = Temperature of the surroundings

h_2 = Radiation constant ($W/m^2.K^4$)

C = Specific heat of the metal. (J/kg K)

All the other values are determined experimentally and substitute in the equation to determine the radiation constant of the metal.

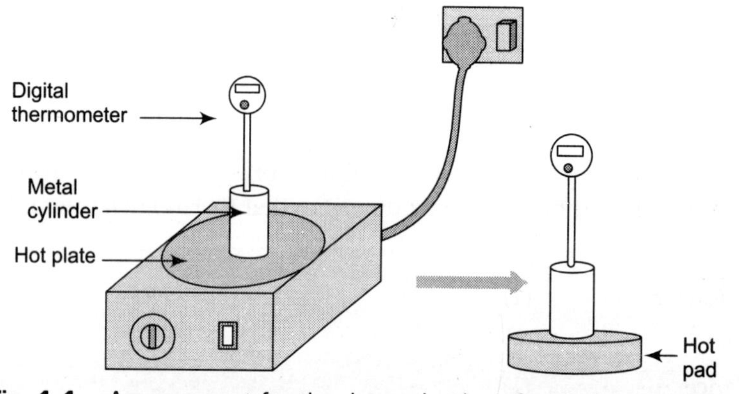

Fig. 1.1: Arrangement for the determination of radiation constant.

PROCEDURE

1. Take a metal cylinder (made up of any metal, for example iron, aluminum, brass, copper of 1 inch diameter, 1.5 inch length which is having drilled hole half way down, so that a thermometer would fit nicely) and clean the surfaces.
2. Find out the surface area and weight of the cylinder (Refer to appendix V).
3. Keep it over a heating plate. (Bunsen burner-tripod stand arrangement can also be used)
4. Insert the digital thermometer to the hole present in the cylinder. (A simple 360 °C thermometer can also be used)
5. Switch on the hot plate.
6. Heat the metal to a maximum temperature. (Depending upon the thermometer maximum)

7. Remove it carefully from the heating source with the help of a tongs.
8. Place it over a hot pad or on the table to minimize the heat loss by conduction.
9. The temperature of the metal cylinder starts to decrease.
10. The temperature readings are noted at an interval of 2 min.
11. Take the readings till the hot body reaches the normal room temperature.
12. The readings are tabulated.
13. Find out the dT/dt.
14. Plot the cooling curve by taking time in min on X axis and temperature on Y axis (Fig. 1.2).
15. Plot another Graph by taking Temperature in °C in X axis and dT/dt in Y axis (Fig. 1.3)
16. Find out the slope of the plot. (Slope = h_1 S/C)
17. Calculate the co-efficient of convection.
18. Substitute the values in the equation to get the radiation constant (h_2).
19. The same procedure can be used for the determination of radiation constant of cylinders/ cubes made up of other metals.

OBSERVATIONS AND CALCULATIONS

Diameter of the cylinder **(d)** = ... cm

= .... m

Radius of the cylinder **(r)** = D/2

= .... m.

Height of the cylinder **(h)** = ...cm

= .... m

Surface area of the cylinder **(S)** = $2\pi r(r+h)$

= ....m^2

Mass of the body = ... gm

= ....kg

Room temperature = ...°C

= ....K

Specific Heat of the metal (°C) = ...J/Kg.K

(Refer to appendix XIII)

Table 1.1: Data for the Construction of Cooling Curve

Sl No	*Time in min*	*Temperature in °C*

Table 1.2: Determination of Co-efficient of Convection

Sl No	*Time (min)*	*Temperature (°C)*	*dT (°C)*	*dT/dt (°C/min)*	*dT/dt (K/sec)*	*h (W/m².K)*
1	2	*a*	-	-		
2	4	*b*	$(a-b)$	$(a-b)/2 = d$	$d + 273.12/60$	
3	6	*c*	$(b-c)$			
4	8					
5	10					
6	12					
7	14					
8	16					
9	18					
10	20					

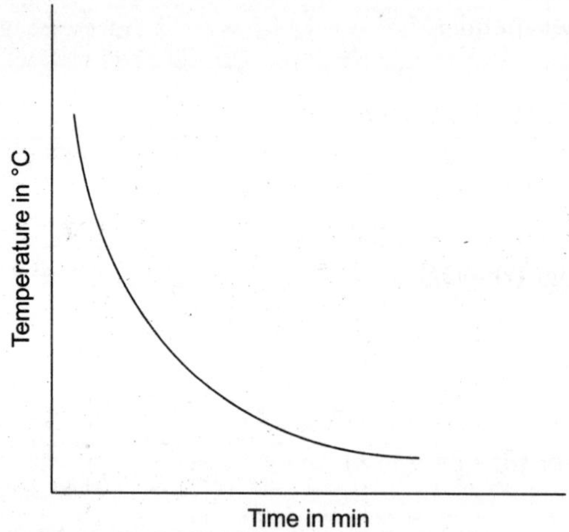

Fig. 1.2: Model graph—Cooling curve.

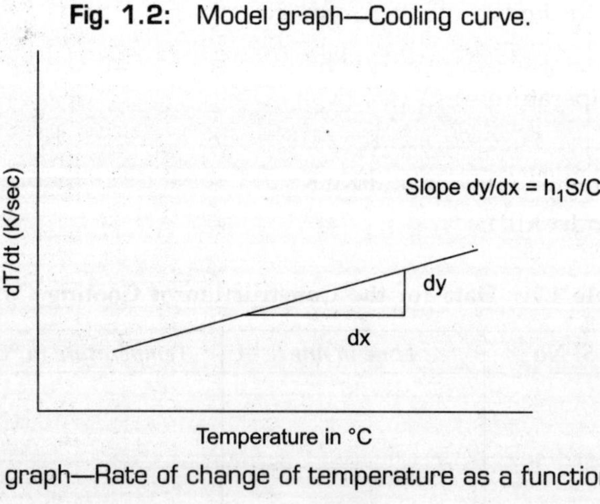

Fig. 1.3: Model graph—Rate of change of temperature as a function of temperature.

Slope of the dT/dt Vs Temperature curve (h_1S/C) = ...

The co-efficient of convectio.. (h_1) = ... W/ m^2.K

The radiation constant (h_2) = ... W/m^2.K^4

Model calculation

$$M\,dT/dt = \frac{h_1S}{C}(T - T_{air}) + \frac{h_2S}{C}(T^4 - T_s^4)$$

$$T = 270\ °C = 543.12K$$

$$T\ air,\ T_s = 30\ °C = 303.12K$$

$$dT/dt = 4.67 \text{ (in K/s from Table 1.2 corresponding to 270 °C)}$$

$$M \times 4.67 = h_1S/C\,(543.12 - 303.12) + \frac{h_2XS}{C}(543.12^4 - 303.12^4)$$

$$h_2 = \ldots\ldots\ldots\ldots\ldots\ldots\ W/m^2.K^4$$

REPORT

The cooling pattern of hot metal body (cylinder) is studied.

The co-efficient of convection (h_1) of the metal cylinder is = ... $W/m^2.K$

The radiation constant (h_2) of the metal cylinder is = ... $W/m^2.K^4$

Experiment 2

Determination of Radiation Constant for Painted and Unpainted Metal Cylinders

AIM

To demonstrate the effect of color on radiation of heat.

To determine the radiation constant of unpainted and painted metal cylinders.

REQUIREMENTS

Digital thermometers	(3)
Metal cylinders	(3)
Painted with white	
Painted with black	
Without painting	
Hot plate	(3)
Weighing balance	(1)
Weighing box	(1)
Tongs	(3)
Hot pad	(3)

PRINCIPLE

A hot surface loses heat (heat is transferred) to its surroundings by the combined modes of convection and radiation (if conduction is negligible). In practice these modes (convection and radiation) are difficult to isolate and therefore an analysis of the combined effects at varying surface temperature provides a meaningful teaching exercise. The rate of heat loss due to combined convection and radiation can be determined by the following equation,

$$M\frac{dT}{dt} = \frac{h_1 S}{C}(T - T_{air}) + \frac{h_2 S}{C}(T^4 - T_s^4) \quad \text{... (Eq. 2.1)}$$

where,

M = Mass of the body (Kg)

dT/dt = Rate of change of temperature (Rate of heat loss) (K/sec)

h_1 = Heat transfer co-efficient for pure convection (W/ m^2.K)

S = Surface area of the hot body (m^2)

T = Temperature of the metal surface (K)

T_{air} = Temperature of the surroundings (K)

T_s = Temperature of the surroundings (K)

h_2 = Radiation constant (W/m^2.K^4

C = Specific heat of the metal (J/kg K)

Substitute all other known parameters in the equation to get the radiation constant.

The present experiment explains the effect of color on the radiation constant and on the cooling pattern.

All warm objects emit electromagnetic radiation. Hotter objects emit more radiation. Objects that emit a lot will quickly cool. Objects that don't emit as much will stay warm longer. Black will absorb more than white, shiny objects will reflect energy and not absorb it. Color also affects emission. For objects near room temperature, which emit radiation in the far infrared part of the spectrum, most are black—that is, they are good absorbers and good emitters. For instance, human skin, regardless of color, is a very good absorber and emitter of far infrared. No matter what color your skin is, you are black in the infrared. The same is true of fabrics and of painted surfaces. All colors of cloths and all colors of paints are black in the infrared; they absorb and emit quite nicely. But silvery metals don't work like this. They reflect visible light and they reflect infrared too. They don't absorb it—and, more importantly, they don't emit it. So, in this experiment, the cylinders (Bare & Painted) will cool at different rates. The bare aluminum cylinder radiates less and cools relatively slower than the white cylinder and black (and the color doesn't matter—it could be any color at all, will radiate more and so it will cool off more quickly).

PROCEDURE

1. Take three aluminum cylinders. (1 inch diameter, 1.5 inch length which are having drilled holes half way down, so that thermometer would fit nicely.)
2. One of the three is left bare; second one is painted with white and the third with black.
3. Turn on the heating mantle.
4. All the cylinders are kept heating on a heating mantle with holes up.
5. Turn on the digital thermometers and insert one into each cylinder.
6. When the metal cylinders are warmed up to a sufficiently high temperature, put off the heating source. (Care should be taken that all the cylinders reach the same high temperature)
7. Remove the metal cylinders from the heating source and place it over a hot pad or on the table.
8. Start noting down the temperature readings at an interval of 2 min.

9. Record the readings.
10. Stop taking the reading when the temperature of the metal cylinders reaches the room temperature.
11. Plot a graph by taking time on X axis and temperature on Y axis (cooling curve)
12. Compare the cooling curves.
13. Find out the dT/dt individually for the three metal cylinders.
14. Plot Graphs by taking Temperature (°C) in X axis and dT/dt (k/sec) in Y axis.
15. Find out the slope of the plots. (Slope = h_1S/C)
16. Calculate the co-efficient of convection for three different conditions. (Grey, white colored, dark grey colored as shown in Fig. 2.1).
17. Substitute the values in the equation to get the radiation constants (h_2).
18. Compare the radiation constants obtained.

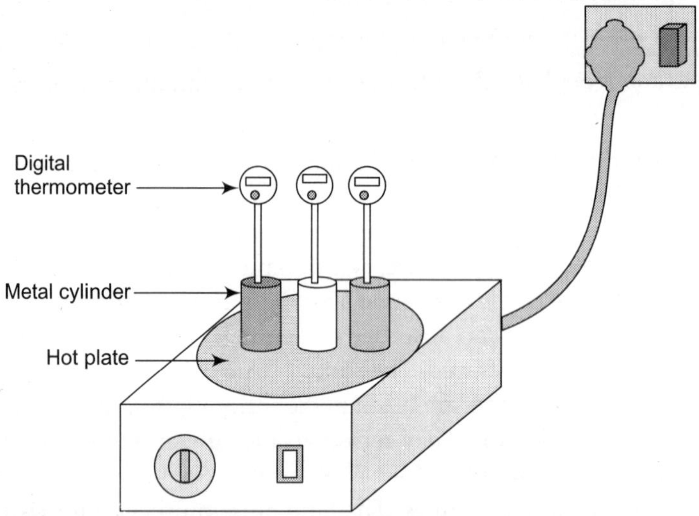

Fig. 2.1: Arrangement for determination of radiation constant.

NOTE: The experiment can also be performed with a glass beaker uniformly heated in hot air oven. The surface area of the beaker can be determined with the help of equation $2 \pi R h + 2 \pi r h$ where R is the external radius and r is the internal radius. (r = R – thickness)

OBSERVATIONS AND CALCULATIONS

Specific Heat of the metal (C) =.........J/Kg.K
(Refer to Appendix XIII)

Table 2.1: Determination of Surface Area of the Cylinders

Condition of the cylinder	*Diameter of the cylinder (D) meter*	*Radius of the cylinder (r) meter*	*Height of the cylinder (h) meter*	*Surface area of the cylinder = (A) m²*	*Weight of the body (M) Kg*
Bare cylinder unpainted					

Cylinder painted white					
Cylinder painted black					

Table 2.2: Cooling Curve Data (Change in Temperature with Time)

Sl No	*Time (min)*	*Temperature Bare cylinder (°C)*	*Temperature cylinder painted white (°C)*	*Temperature Cylinder painted black (°C)*
1				
2				
3				
4				
5				
6				
7				
8				
9				
10				

Table 2.3: Determination of Co-efficient of Convection

Sl No	*Time in min*	*Temperature (°C)*			*dT (°C)*			*dT/dt (°C/min)*			*dT/dt (K/sec)*			*h .(W/ m².K)*		
		Bare	**White**	**Black**	**Bare**	**White**	**Black**	**Bare**	**White**	**Black**	**Bare**	**White**	**Black**	**Bare**	**White**	**Black**
1																
2																
3																
4																
5																
6																
7																
8																
9																
10																

(Refer to previous experiment for model calculation)

Slope of the dT/dt Vs Temperature curve (h_1**S/C**) = ...

The co-efficient of convection (h_1) = ... W/ m^2.K

$$M\,dT/dt = \frac{h_1 S}{C}(T - T_{air}) + \frac{h_2 S}{C}(T^4 - T_s^4)$$

The radiation constant **(h_2)** = ... W/m^2.K^4

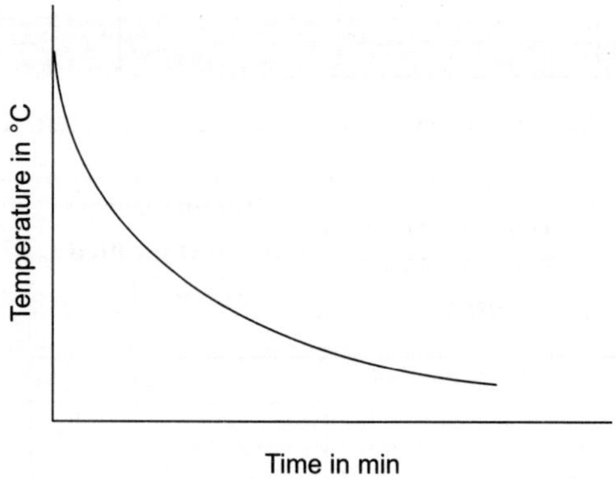

Fig. 2.2: Model graph—Cooling curve.

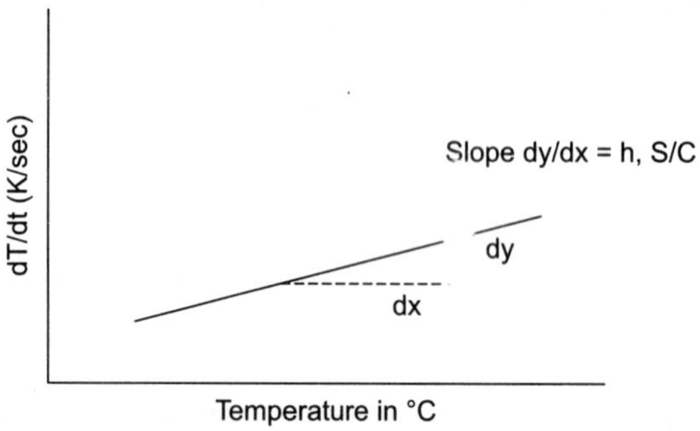

Fig. 2.3: dT/dt VS Temperature.

REPORT

The effect of color on radiation is studied.

The rate of cooling of .... cylinder > .... cylinder > ....cylinder

The radiation constant of the

Bare metal cylinder = ... W/m^2.K^4

White painted cylinder = ... W/m^2.K^4

Black painted cylinder = ... W/m^2.K^4.

EXPERIMENT 3

DETERMINATION OF OVERALL HEAT TRANSFER CO-EFFICIENT

AIM

To determine the overall heat transfer co-efficient based on the inner and outer surface of a condenser.

REQUIREMENTS

Round bottom flask	(1)
Condenser	(1)
Conical flask	(1)
Heating source (burner or mantle)	(1)
Burette stands	(1)
Beaker 250 ml	(1)
Thermometer	(1)
Stop watch	(1)
Plastic pipe with adapter	(2)
Measuring cylinder	(1)

PRINCIPLE

The overall heat transfer co-efficient **U** is a measure of the overall ability of a series of conductive and convective barriers to transfer heat. It is commonly applied to the calculation of heat transfer in heat exchangers.

A condenser used in the laboratories is considered as a simple heat exchanger and the demonstration of overall heat transfer co-efficient can be done with the help of this. For a condenser, the rate of heat transfer can be determined by simple equation based on the flow rate of the cold liquid

$$q = m_c c_{pc} (T_{cb} - T_{ca}) \quad \text{... (Eq. 3.1)}$$

where,

q = Rate of heat transfer

m_c = Mass flow rate of cold water (liter/sec)

c_{pc} = Specific heat of cold water (J/kg)

T_{cb} = Temperature of the leaving cold water (°C)

T_{ca} = Temperature of the entering cold water (°C)

Rate of heat transfer, based on the area of heating surface (tube) can be determined as follows,

$$dq/da = U\Delta T \quad \text{... (Eq. 3.2)}$$

dq/da = Local flux, ΔT = Over all local temperature difference ($T_h - T_c$, where T_h is the average temperature of the hot fluid, T_c is the average temperature of the cold fluid)

U = Local overall heat transfer co-efficient

Since ΔT can vary considerably from point to point along the tube, the heat flux is proportional to ΔT. To apply this equation for the total surface area, the equation must be integrated. This can be done with some simple assumption that, the overall co-efficient U is a constant, the specific heat of hot and cold fluids are constant, the flow is steady in both parallel or counter current. After integration and proper arrangement the equation becomes,

$$q = \frac{UA\ \Delta T_2 - \Delta T_2 - \Delta T_1}{\ln(\Delta T_2/\Delta T_1)} = UA\ \Delta T_L \quad \text{... (Eq. 3.3)}$$

where,

q = Rate of heat transfer

U = Overall heat transfer co-efficient

A = The total surface area of heat transfer

$$\Delta T_L = \frac{\Delta T_2 - \Delta T_1}{\ln(\Delta T_2/\Delta T_1)} \quad \text{... (Eq. 3.4)}$$

When ΔT_1 and ΔT_2 are nearly equal the arithmetical average can be used instead of ΔT_L

Thus the overall heat transfer co-efficient can be written as,

$$U = \frac{q}{A\Delta T_L} \quad \text{... (Eq. 3.5)}$$

In the condenser the actual heating surface is glass tube of certain thickness, so the heat transfer is based on either the inner surface area or the outer surface area.

So,

$$q = U_i A_i \Delta T_L \quad \text{... (Eq. 3.6)}$$

$$q = U_o A_o \Delta T_L \quad \text{... (Eq. 3.7)}$$

From this we can determine the overall heat transfer co-efficient based on inner and outer surface area

$$U_i = q/A_i \Delta T_L \quad \text{... (Eq. 3.8)}$$

$$\mathbf{U_o = q/A_o \Delta T_L} \qquad \text{... (Eq. 3.9)}$$

Since ΔT_L and q are independent of the choice of area we can write,

$$U_oA_o = U_iA_i$$

So,

$U_o/U_i \quad A_i/A_o \quad D_i/D_o$

Thus,

$$\mathbf{U_o = U_iA_i / A_o} \qquad \text{... (Eq. 3.10)}$$

$$\mathbf{U_i = U_o A_o / A_i} \qquad \text{... (Eq. 3.11)}$$

In plate type heat exchanger the area on both sides are same so there is only one value of U.

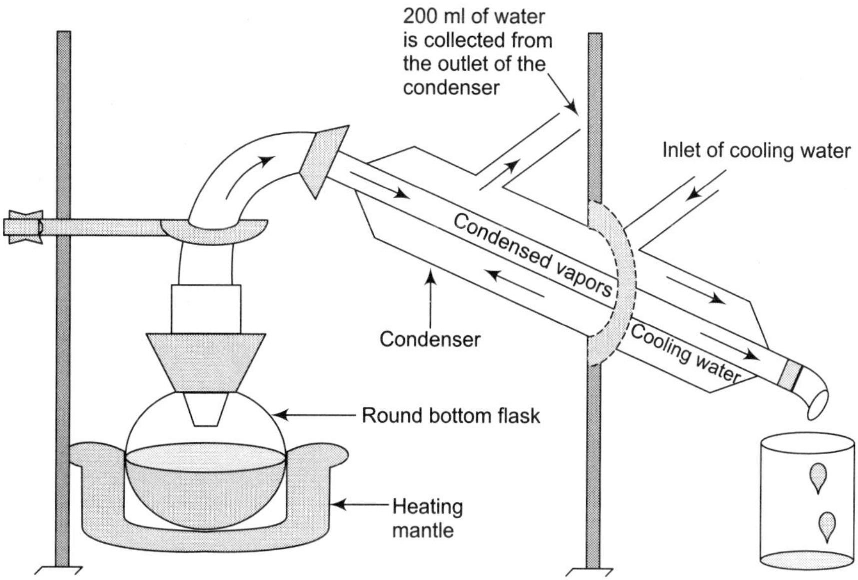

Fig. 3.1: Assemble for the determination of overall heat transfer co-efficient.

PROCEDURE

1. Assemble the condenser to round bottom flask containing 250 ml of water.
2. Inlet of the condenser is attached to the tap, and the outlet is placed to a 250 ml beaker.
3. Arrange the assembly in such a way that it can be heated with a suitable heating medium. (either burner or mantle) (Temperature 100 °C).
4. Steam is generated by heating.
5. When steam is steady through the condenser, start collecting the liquid from the outlet of the condenser and time is noted.
6. Collect 200 ml of water and corresponding time to collect 200 ml is also noted.
7. The experiment is repeated for different water velocity by adjusting the knob of the tap.
8. For each velocity inlet and outlet temperature of the cold water is noted (Water circulating through the condenser)
9. The inner and outer diameters of the condenser tube are determined.
10. The length of the inner tube of the condenser is determined.

11. Find out the mass flow rate.
12. Find out the heat transfer and overall heat transfer co-efficient.

Note : In this present experiment, we are not bothering about the amount of the water collected in the receiver. We are only dealing with the water circulating through the condenser.

OBSERVATIONS AND CALCULATIONS

Outer diameter of the inner tube (D_o) = ———————— cm

Inner diameter of the inner tube (D_i) = ———————— cm

Length of the condenser (h_o) = ———————— cm

Volume of water collected = 200 ml

Specific heat of water (C_{pc}) = 4.18×10^3 J/kg

Table 3.1: Determination of ΔT_L

Sl No	Time taken to collect 200ml (sec)	Inlet temperature T_1 (°C)	Outlet temperature T_2 (°C)	$T_2 - T_1$ ΔT	Δt_1 $(100 - T_1)$	Δt_2 $(100 - T_2)$	$\Delta t_L = \dfrac{\Delta t_1 - \Delta t_2}{\ln \Delta t_1/\Delta t_1}$
1.							
2.							
3.							

Table 3.2: Determination of Rate of Heat Transfer and Overall Heat Transfer Co-efficient

Sl No	*Mass flow rate* $m_c = \dfrac{\rho \times Volume}{Time \times 1000}$	$Q = m_c c_{pc} \Delta t$	U_o	U_i
1				
2				
3				
4				
5				

$$\text{Mass flow rate } (m_c) = \frac{\text{Density of water} \times \text{Volume}}{\text{Time for collecting 200 ml} \times 1000}$$

$$= \text{------------} \text{ liter/sec}$$

$$\text{Outer surface area of the inner tube } (\mathbf{A}_o) = \Pi \, D_o \, h_o$$

$$= \text{---------} \text{ cm}^2$$

$$\text{Inner surface area of the inner tube } (\mathbf{A}_i) = \Pi \, D_i \, h_o$$

$$= \text{------------} \text{ cm}^2$$

$$\text{Rate of heat transfer } (\mathbf{q}) = m_c \, c_{pc} \Delta t$$

Overall heat transfer co-efficient based on outer surface area of the inner tube of condenser

$$= \mathbf{U_o} = \frac{q}{A_o \, \Delta T_L}$$

Overall heat transfer co-efficient based on inner surface area of the inner tube of condenser

$$= \mathbf{U}_i = \frac{q}{A_i \, \Delta T_L}$$

REPORT

The overall heat transfer co-efficient based on inner and outer surface area is calculated and reported

U_o = ------------------

U_i = ------------------

EXPERIMENT 4

DETERMINATION OF TYPE OF FLOW BY REYNOLD'S APPARATUS

AIM

To determine the Reynold's number by Reynold's apparatus.
To demonstrate the type of flow (laminar, turbulent, or transition zone).
To determine the critical velocity of flow of a liquid.

REQUIREMENTS

Reynold's apparatus	(1)
Pump	(1)
Water tank	(1)
Colored solution ($KMnO_4$)	

PRINCIPLE

Fluid can flow through a pipe in two different ways. At low flow rates the pressure drop in the fluid increases directly with the fluid velocity and at high rates it increases much more rapidly. At low velocity the fluid will travel as layers and there is no mixing of the layers of the fluid. This type of flow is called as laminar flow. At high velocity this pattern of flow is disturbed and there will be a complete mixing of the layers and production of turbulence. This type of flow can be called as turbulent flow.

The distinction between the two flows can be demonstrated with the help of Reynold's experiment. Reynold's experiment uses a transparent glass tube with central nozzle for the introduction of colored solution. Water is allowed to pass through the glass tube at low velocity and through the nozzle colored solution is also introduced. At low velocity it is found that the colored solution is traveling as a straight line as there is no mixing of the layers. By slowly increasing the velocity of the flow a particular velocity is reached where the thread of color become wavy and gradually disappeared. This velocity of flow is called as critical velocity or transition velocity. At this particular velocity the flow is neither laminar nor turbulent. The critical velocity is dependent on the diameter of the tube, viscosity of the fluid, density of the fluid and average linear velocity of the liquid.

Reynold's equation combines all this factors affecting the flow pattern and can be written as,

$$\boxed{Re = \rho V d / \mu} \quad \text{... (Eq. 4.1)}$$

where,

Re = Reynold's number
ρ = Density of the fluid in Kg/m^3
μ = Viscosity of the fluid in Ns/m^2
V = Average velocity of the fluid flow in m/sec
d = Internal diameter of the glass tube in m

If the Reynold's number is

Below 2000, the flow is considered as laminar flow

Between 2000–4000 flow is in transition zone

Above 4000 turbulent flow

The corresponding velocity can be determined by

$$V = Q/A \quad \text{... (Eq. 4.2)}$$

where,

$Q = V/t$ (m^2/s)
A = Area of the glass tube (m^2)

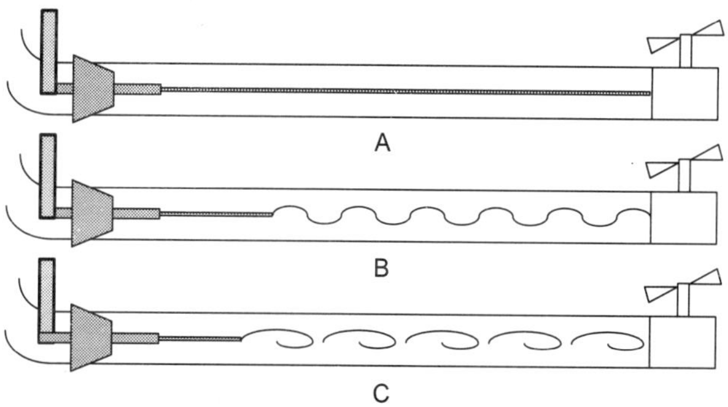

(a) Showing stream line of the dye solution (laminar flow)
(b) Transition zone (c) Turbulent flow

Fig. 4.1: Type of flow with flow pattern.

PROCEDURE

Description of the Apparatus

The apparatus consist of a glass tube with a wide mouth entrance connected to water tank. At the opposite end a control valve is provided to change the flow rate. The rate of flow can be determined with the help of a measuring cylinder and stop watch. At the centre of the wide mouth end of the glass tube a capillary tube is introduced. Through this tube dye solution is introduced.

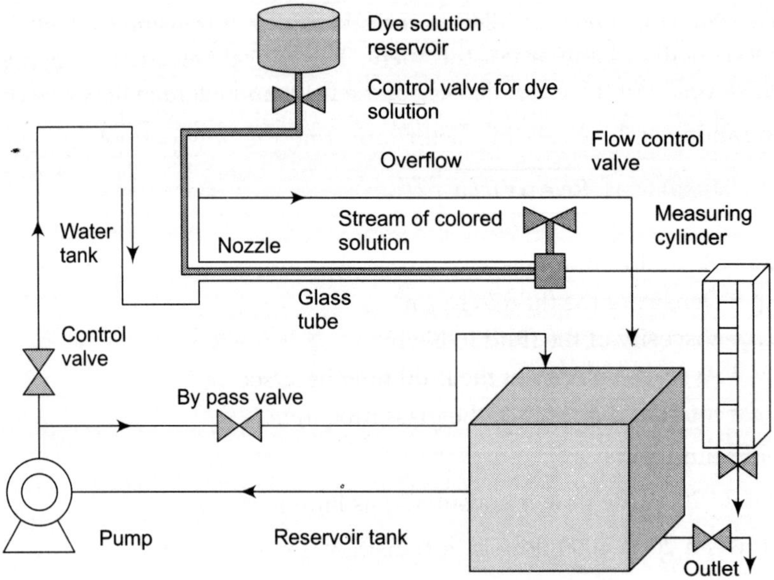

Fig. 4.2: Assemble of Reynold's apparatus.

OPERATING PROCEDURE

1. Clean the apparatus make all the parts of the apparatus dust free.
2. Close the drain valves provided.
3. Fill the reservoir tank ¾th with clean water.
4. Prepare a dye solution ($KMnO_4$) 1 liter in pure water. Close the flow control valve for the dye solution.
5. Put this solution on the dye vessel after removing all the solid particles from it.
6. Open the flow control valve and the bypass valve.
7. Switch on the power supply.
8. Switch on the pump.
9. The water is filled in the water tank, and the overflow is maintained at minimum.
10. Allow the water to pass through the glass tube from the water tank at minimum flow rate by regulating the valve at the end of the glass tube.
11. Adjust the flow of the dye solution through the capillary tube so as to get a narrow fine colored line of the dye solution at the centre of the glass tube.
12. Determine the flow rate with the help of measuring cylinder and stop clock.
13. Now slowly increase the flow through the glass tube with the help of flow control valve present at the end of the glass tube.
14. Now the clear colored solution starts to travel in a wavy form (Upper critical velocity).
15. Find out the discharge and the flow rate.
16. Further increase the flow rate so that colored solution is completely diffusing in to the water (Turbulent flow).
17. Determine the discharge and the flow rate.

18. Decrease the flow of the liquid so as to get the flow in such a way that the colored solution travels in wavy form. Find out the discharge and flow rate.
19. Again decrease the flow so that the wavy fashion of the flow of the colored solution starts to change in to straight line. Find out the flow rate (lower critical velocity).
20. Finally keep the flow in such a way that the colored solution is moving in a straight line without any disturbance.
21. After taking all the readings switch off the pump and the power supply.

OBSERVATIONS AND CALCULATIONS

Internal diameter of the glass tube = ... mm

= ....m

Area of the glass tube = ...m^2

Kinematic viscosity of water **(υ)** = at 0 °C = 1.788×10^{-6} m^2/s

= at 20 °C = 1.006×10^{-6} m^2/s

Table 4.1: Determination of Type of Flow and Reynold's Number

Sl No	*Type of flow observed*	*Volume collected (ml)*	*t (sec)*	*V(m/sec)*	*Q (m^3/s)*	*Reynold's number (Re)*
1						
2						
3						
4						
5						
6						

To convert liter to m^3 multiplied with 0.001

Reynold's number **(Re)** = $\rho V d / \mu$

where,

ρ = Density of the fluid in kg/m^3

μ = Viscosity of the fluid in Ns/m^2

V = Average velocity of the fluid flow in m/sec

v = Volume of water collected m^3

t = Time taken to collect v m^3 of water

A = Area of the glass tube

μ/ρ = Kinematic viscosity m^2/s

Discharge of the fluid **Q (V/*t*)** = ...m^3/s

Velocity of the fluid **(V)** = Q/A

= ....m/sec

REPORT

Different types of flow pattern is demonstrated and studied

Reynold's number for

Laminar flow =...............

Transition zone =.............

Turbulent flow =.............

The critical velocity is found to be .... m/sec.

EXPERIMENT 5

DETERMINATION OF FLOW RATE BY DIRECT WEIGHING OR MEASURING

AIM

To determine the flow of fluid by direct weighing or measuring.

REQUIREMENTS

Long tube connected to a water inlet	(1)
Measuring tank	(1)
Measuring cylinder of capacity (1000 ml)	(1)
Weighing balance	(1)
Stopwatch	(1)

PRINCIPLE

One of the methods of determination of flow rate is by direct weighing or measuring. In this method no other parameters are considered for calculating the flow rate. The amount (volume) of the liquid coming out from the pipe up to a particular time is determined and from the value, the volume of liquid passing through the pipe per min is calculated. Same way the weight of the liquid collected for different time Interval is determined and the flow rate in weight per min or sec is found out.

PROCEDURE

1. Arrange a pipe line (straight) of known length.
2. Attach one end of the pipe line to the water tap with suitable arrangements.
3. The opposite end of the pipe is fixed into a measuring cylinder.
4. Allow to pass the water at slow rate through the pipe by adjusting the knob of the tap.
5. Start the stopwatch, when the first drop of the water is collected the measuring cylinder.

6. Find out the time taken to fill the 500 ml of water in the measuring cylinder.
7. Note the reading.
8. Find out the volume of the liquid collected at 1 min.
9. Discard the water in the measuring cylinder.
10. Increase the flow through the pipe by adjusting the knob on the tap.
11. The steps 5–8 are repeated.

OBSERVATIONS AND CALCULATIONS

Table 5.1: Determination of Flow Rate

Sl No	*Time taken to collect 500 ml of water (t) (min)*	*Flow rate (ml/min) 500/t*
1		
2		
3		
4		
5		

REPORT

The flow rates of the water through the pipe are determined and tabulated.

EXPERIMENT 6

DETERMINATION OF PRESSURE DIFFERENCE DURING LIQUID FLOW USING SIMPLE MANOMETER

AIM

To determine the pressure difference in a flow of liquid through a pipe with the help of simple U-tube manometer.

REQUIREMENTS

Orifices meter or venturi meter assemble	(1)
Pump	(1)
Manometer	(1)
Water	

PRINCIPLE

A manometer, when used properly, is a very accurate instrument for the determination of pressure difference during fluid flow. The U-type manometer is a primary standard due to its inherent accuracy and simplicity of operation. The manometer has no moving parts subject to wear, age, or fatigue. U-tube Manometer consists of two limbs which are attached to the pipe through which the fluid is flowing. Both the limbs are filled with mercury in equal quantity. If there is no pressure difference both the limbs show same height of the mercury level on the scale. If any pressure difference occurs between the limbs, mercury level in the limb will differs and shows increased or decreased scale readings. The pressure difference can be determined by the equation,

$$\boxed{\Delta p = P_2 - P_1 = h\,(\rho_A - \rho_B)g} \quad \text{...(Eq. 6.1)}$$

where,

Δp = Differential pressure

P_1 = Pressure applied to one liquid surface

P_2 = Pressure applied to the other liquid surface
ρ_A = Density of the mercury 13.5951(g/cm^3)
ρ_B = Density of water (g/cc)
g = Acceleration of gravity (9.8 m/s^2)
H = Height of the liquid column (m)

So the equation can be reduced into,

$$\Delta p = h \times 12.6 \times g \quad \text{... (Eq. 6.2)}$$

PROCEDURE

(Refer to Experiment Number 12 and Fig. 12.3)

1. Clean the apparatus and tank properly to avoid any interruption in the experiment.
2. Check the power supply.
3. Close the drain valve (V_6).
4. Fill water in the tank (¾ th).
5. Close all the valves except By pass valve V_2.
6. Check the level of mercury in manometer tube.
7. Close all pressure tapes of manometer connected to the orifice meter. (P_1 & P_2)
8. Switch on the power supply and the pump.
9. Operate the flow control valve to allow the flow through the orifice meter.
10. Open the pressure taps of the manometer slowly, care should be taken to avoid the flow of liquid to the manometer.
11. Open the air release valve to release the air.
12. When there is no air in the manometer close the air release valve.
13. Note down the manometer reading.
14. Increase the pressure of flow by operating the pressure control valve.
15. Steps 10–13 are repeated.
16. Repeat this for 4–5 different flow rates.
17. Finally when all the reading is recorded open the Bypass valve fully. Then close the flow control valve.
18. Close all the manometer pressure taps.
19. Now switch of the pump and the power supply.
20. Determine the pressure difference for different flow rates.

Instructions for Noting the Readings

For consistent results, it is necessary that the fluid meniscus always be observed in the same way. If mercury is used, it will show a convex meniscus. So the fluid level should be observed from the upper most point. For all other indicating fluids a concave meniscus forms. The reading in this case should be observed from the lowest point of the meniscus.

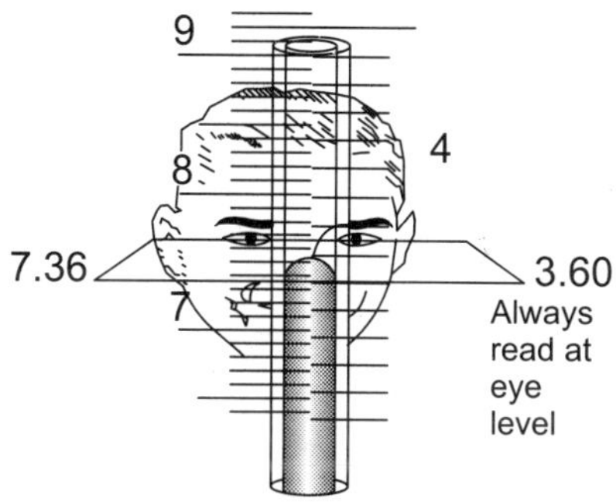

Fig. 6.1: Showing the correct observation of mercury levels.

OBSERVATIONS AND CALCULATIONS

Table 6.1: Determination of Pressure Difference During Fluid Flow

Sl No	*Height of mercury left limb (cm) h_1*	*Height of mercury right limb (cm) h_2*	*h(cm) $h_1 - h_2$*	*h(m)*	*$H = h\ (m) \times 12.6$*	*$\Delta p = h \times 1000 \times 9.8$*
1						
2						
3						
4						
5						
6						

Height of the liquid column in the manometer $\boldsymbol{h} = h_1 - h_2$

$= \ldots\ldots\ldots..\text{cm}$

$= \ldots\ldots\ldots...\text{m}$

$\mathbf{H} = 12.6^* \times h\ (\text{m})$

$= \ldots\ldots\ldots\ldots$

$\boldsymbol{\Delta p} = h \times 12.6 \times g$

where,

$\text{H} = h \times 12.6$

$\Delta p = \text{H} \times g \times 1000$

$= \ldots\ldots\ldots\ldots\ldots \text{N/m}^2$

$g = 9.8 \text{ m/sec}^2$

*Note : 12.6 is the difference in the densities of mercury and water. Density of mercury 13.6 gm/ cm^3 and density of water 1 gm /cm^3.

REPORT

Determined the pressure differences at different flow rate with the help of simple manometer.

EXPERIMENT 7

DETERMINATION OF PRESSURE OF GAS FLOW BY U-TUBE MANOMETER

AIM

To determine the pressure of the gas flow using U-tube manometer.

REQUIREMENTS

U-tube manometer	(1)
Pipe connections	(2)
Gas source	(1)
Acetone	
Mercury	

PRINCIPLE

Simple U-tube Manometer consists of a U shape transparent tube of height 10 mm or more, U-tube manometer can measure both positive and negative pressures. One of the limbs of U-tube manometer is longer than the other and is kept open to the atmosphere. This is called as open limb. The other limb is connected to the pressure tapping. The mercury level (bottom of the concave meniscus of the mercury) in each limb is read and the difference is calculated. This is called as deflection. This will be the pressure of the flow of the gas. With the help of conversion factors, the pressure in mm Hg can be converted into other units (Refer to appendix I & II).

The open-tube manometer is used to measure the pressure of a gas in a container. The pressure of the gas is determined by finding out the h (the difference in mercury levels) in units of torr or mmHg. During the experiment one of the limbs is connected to the gas source and the other limb kept open. The Atmospheric pressure pushes the mercury from one direction, and the gas flow from the container from the other direction. So for the determination of the gas pressure during flow can be determined using the equations,

If $\mathbf{P_{gas} > P_{atm}}$

Gas pressure = atmospheric pressure + *h* ... (Eq. 7.1)

If $\mathbf{P_{gas} < P_{atm}}$

Gas pressure = atmospheric pressure – *h* ... (Eq. 7.2)

where,

h = Difference in the mercury levels (m)

A closed-tube manometer looks similar to regular manometers except that the end that's open to the atmospheric pressure in a regular manometer is sealed and contains vacuum. In these systems, the difference in mercury levels (in mmHg) is equal to the pressure in torr. A number of liquids can be used as manometeric fluid, but it should fulfill the following requirements:

1. It should not mix with the flowing fluid.
2. It should not chemically react with the flowing fluid or the pipe wall material.
3. It should have higher density than the flowing fluid.
4. It should not stick (adhere) to the tube.
5. Its boiling point should be much higher than the atmospheric temperature.
6. It should be clearly visible.

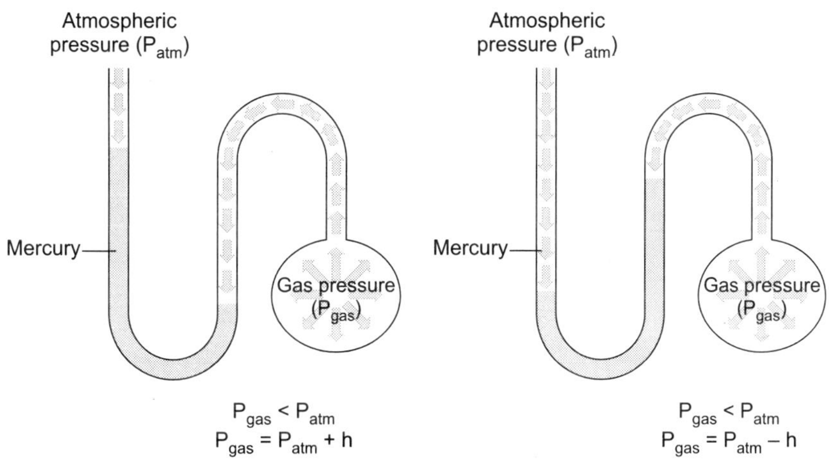

Fig. 7.1: Mercury levels in the manometer.

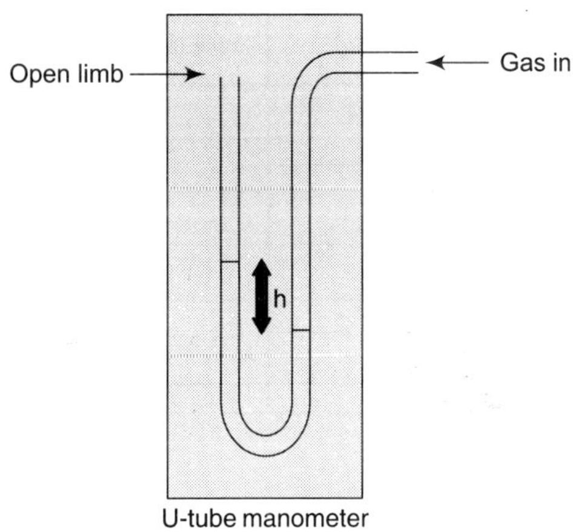

Fig. 7.2: Arrangement of manometer for the determination of Pressure of gas flow.

PROCEDURE

1. Clean the U-tube with water and rinse it with acetone.
2. Fill the mercury in the U-tube with the help of suitable means to avoid spillage of mercury.
3. Check the scale and the mercury level in the limbs. (Mercury should show equal height on both the limbs)
4. The tube from one arm of the U-tube is connected to the gas source. (It can be a gas cylinder or burner gas connection, vacuum pump, cycle pump etc)
5. Open the knob of the gas source.
6. Find out the difference in the mercury level of the limbs.
7. Pressure difference is recorded in terms of mm of Hg.
8. Repeat the procedure for two more times to get concordant results.
9. Pressure difference can be determined for different flow rates of gas (by adjusting the knob of the gas source).

Note:

1. Care should be taken if dealing with gas. Ensure that the experimental arrangement is away from the heating source. Experiment should be completed within minimum time in order to avoid the inhalation of gas. If vacuum pump is used care should be taken to avoid the drainage of the mercury along with the vacuum stream.
2. Manometer assemble along with pipe connections are available in the market.

OBSERVATIONS AND CALCULATIONS

Atmospheric pressure = 760 mmHg
Mercury level in the limb (h_1) = ...mmHg
Mercury level in the limb (h_2) = ...mmHg
Difference in the levels in the limbs (h) = ...mmHg
h = ...mmHg
Gas pressure = ...mmHg
Gas Pressure = ...Pa (N/m^2)

Table 7.1: Determination of Pressure of Gas Flow

Sl No	*Height of mercury on the left limb h_1 (mm Hg)*	*Height of mercury on the right limb h_2 (mm Hg)*	*Difference in height of mercury (h) (mmHg)*	*Pressure of gas flow (mmHg)*	*Pressure of the gas flow (Pa)*
1					
2					
3					
4					
5					
6					

REPORT

The pressure of the gas flow at different flow rates is determined.

EXPERIMENT 8

DETERMINATION OF PRESSURE OF GAS FLOW BY INCLINED MANOMETER

AIM

To determine the pressure of flow of gas by inclined manometer.

To find out the gas pressure at different gas flow rate.

REQUIREMENTS

Inclined manometer	(1)
Pipe connections	(1)
Gas source	(1)
Scale	(1)
Protractor	(1)

PRINCIPLE

Inclined manometer is used to determine small differences in pressure during fluid flow. In this type of manometer, one of the legs is inclined at angle α and the liquid in the limb should move a considerable distance for a small difference in pressure. The inclined manometer or slant manometer is an enlarged leg manometer with its measuring leg inclined to the vertical axis by some angle. The inclination is done to expand the scale & thereby to increase the sensitivity. Commonly used Manometric liquids in inclined manometer are:

1. **Mercury:** It has low freezing point – 38 °F & high boiling point 675 °F but it corrodes many metal & it is poisonous & expensive.
2. **Water with coloring agents:** Coloring agents reduce surface tension of pure water that reduces capillary effect in manometer.
3. Benzene, kerosene, CCl_4, toluene etc.

Among this, mercury is preferred for its high density (13.5951 g/cm^3) and low vapour pressure. For low pressure differences (well above the vapour pressure of water) water is commonly used (and "inches of water" is the common pressure unit). Inclined manometer is also an open type of manometer so the criteria for the U-tube manometer is applicable to inclined manometer also.

The pressure of the gas is given by *h* (the difference in mercury levels) in units of torr or mmHg. The Atmospheric pressure pushes on the mercury from one direction, and the gas flow from the container from the other direction. So for the determination of the gas pressure during flow of fluids can be determined by the equations,

If $P_{gas} > P_{atm}$

$$\textbf{Gas pressure} = \textbf{atmospheric pressure} + h \quad \text{... (Eq. 8.1)}$$

If $P_{gas} < P_{atm}$

$$\textbf{Gas pressure} = \textbf{atmospheric pressure} - h \quad \text{... (Eq. 8.2)}$$

where,

The difference in the mercury level in the limbs can be determined by,

$$\boxed{h = R \sin \alpha} \quad \text{... (Eq. 8.3)}$$

where,

h = Difference in the mercury levels (mmHg)

R = Difference in the readings in the inclined limb (mmHg)

α = Angle of inclination (°)

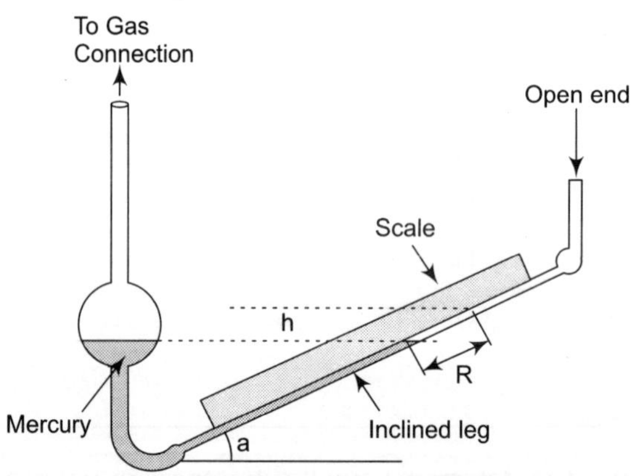

Fig. 8.1: Inclined manometer.

PROCEDURE

1. Clean the inclined manometer with water and rinse it with acetone.
2. Fill the mercury in the manometer with the help of suitable means to avoid spillage of mercury.
3. Check the scale and the mercury level in the limbs.
4. Find out the angle of inclination with the help of scale and protractor.
5. The tube from Vertical arm of the inclined manometer is connected to the gas source. (It can be a gas cylinder or burner gas connection, vacuum pump, cycle pump etc).

6. Open the knob of the gas source.
7. Find out the difference in the mercury level of the limbs R as shown in Fig. 8.1.
8. Pressure difference (h) is determined in terms of mm of Hg.
9. Repeat the procedure for two more times to get concordant results.
10. The experiment can be repeated for different gas velocity.
11. The observations are tabulated.

OBSERVATIONS AND CALCULATIONS

Table 8.1: Determination of the Gas Pressure at Different Flow Rate

Sl No	*Sin α*	*Mercury Level in the vertical limb (mmHg)*				*Mercury level in the inclined limb (mmHg)*				*R(mmHg)*	*h(mmHg)*	*Gas pressure (mmHg)*	*Gas pressure (Pa)*
		1	2	3	av	1	2	3	av				
1													
2													
3													
4													

Atmospheric pressure = 760 mmHg

Angle of inclination **(α)** = ...

sin α = ...

Mercury level in the vertical limb = ...mmHg

Mercury level in the inclined limb = ...mmHg

Difference in the levels in the limbs **(R)** = ...mmHg

h = R sin α

= .... mmHg

Gas pressure (Eq. 8.1 & 8.2) = ...mmHg

Gas Pressure = ...Pa(N/m^2).

REPORT

The pressure of gas flow is determined with the help of inclined manometer and studied the change in pressure at different flow rates.

EXPERIMENT 9

DETERMINATION OF FLOW OF FLUID—ROTAMETER

AIM

To determine the flow of fluid through a pipe using Rotameter.

REQUIREMENTS

Rotameter	(1)
Pipe connection	(2)
Measuring cylinder	(1)
Water	

PRINCIPLE

Rotameter consists of a tapered glass tube with the smallest diameter at the bottom. The tube contains a freely moving Float/plummet which rests on a stop at the base of the tube. When the fluid is flowing, the float rises until its weight is balanced by the up thrust of the fluid and reaches a position of equilibrium, this position then indicates the rate of flow. Floats can be of many shapes, but spheres and ellipsoids being the most common. The float is shaped so that it rotates axially as the fluid passes. Readings are usually taken from the top of the float.

The rotameter's operation is based on the variable area principle: Fluid flow raises the float in a tapered tube, increasing the area for passage of the fluid. The greater the flow, the higher the float is raised. The height of the float is directly proportional to the flow rate. With liquids, the float is raised by a combination of the buoyancy of the liquid and the velocity head of the fluid. With gases, buoyancy is negligible, and the float responds to the velocity head alone.

The float moves up or down in the tube in proportion to the fluid flow rate and the annular area between the float and the tube wall. The float reaches a stable position in the tube when the upward force exerted by the flowing fluid equals the downward gravitational force exerted by the weight of the float. A change in flow rate upsets this balance of forces. The float then moves up or down, changing the annular area until it again reaches a position where the forces are in equilibrium. To satisfy the force equation, the rotameter float assumes a distinct position for every constant flow rate. It is important that the rotameter must be vertically oriented and mounted. (The float position is gravity dependent).

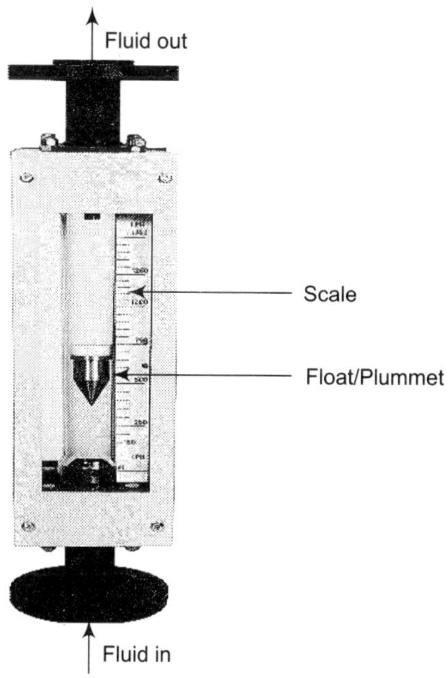

Fig. 9.1: Rotameter.

PROCEDURE

1. Clean the apparatus with fresh water.
2. Arrange the assembly as shown in Fig. 9.1. Connect the lower opening of the Rotameter to the water source (tap) and upper opening to a measuring jar (1000 ml capacity) with the help of pipes provided.
3. Check the position of the float/plummet before starting the experiment.
4. Slightly open the tap so that water flows to the rotameter at lowest rate possible.
5. Stop clock is started when the first drop of water starts to collect in the measuring jar.
6. Wait till 1000 ml of water is collected in the jar.
7. Note down the time required for collecting 1000 ml of water in the measuring jar.
8. Note down the height of the plummet in the rotameter.
9. This procedure is repeated for different flow rates.
10. Try up to the flow rate by which the plummet reaches its maximum possible height.
11. Plot a graph by taking flow rate on X axis and height of the plummet on Y axis.
12. Make a strip of length as that of the rotameter tube and a width of 2-3 cm with a graph paper and stick on the rotameter tube.
13. Mark the flow rate (ml/min) with the help of the plot flow rate Vs height of the plummet (mm).
14. Now open the tap at any flow rate (unknown) and find out the flow rate from the calibrated values.

OBSERVATIONS AND CALCULATIONS

Table 9.1: Determination of Flow Rate

Sl No	*Time for collecting 1000 ml water (min)*	*Height of the plummet (cm or inch)*	*Flow rate (ml/Min)*
1			
2			
3			
4			
5			
6			
7			
8			
9			
10			

REPORT

The rotameter is calibrated for different flow rate.

The flow rate of the unknown velocity = ... ml/min

= .... cc/sec.

EXPERIMENT 10

DETERMINATION OF FRICTIONAL PRESSURE LOSS DURING FLUID FLOW THROUGH A PIPE

AIM

To determine the pressure loss due to friction during the flow through a pipe.

To determine the relative roughness of the pipe using the Moody's diagram.

To determine the fanning frictional factor.

REQUIREMENTS

Long PVC pipe connected to a tap/long PVC pipe connected to a pump with regulating valve (1)

Collecting tank (measuring tank)	(1)
Stopwatch	(1)
Long scale	(1)
Vernier calipers	(1)

PRINCIPLE

As the fluid flows through a pipe from a point A to B the pressure decreases due to frictional loss between the flowing liquid and the pipe. The extent of pressure loss designated in feet depends on many factors. These factors include liquid flow rate, liquid specific gravity and viscosity, diameter of pipe (inside), length, and internal conditions of the pipe(smooth -rough). In order to find out the frictional loss, the fanning's frictional equation is used,

$$\boxed{h_f = f_f\,(L/R_h) \times (V^2/2g)} \qquad \text{... (Eq. 10.1)}$$

where,

h_f = Frictional pressure loss in (m) of liquid head

f_f = Fanning frictional factor

L = Length of pipe (m)

R_h = Hydraulic radius of pipe (m)

V = Velocity of fluid (m/s)

g = Acceleration due to gravity (m/s^2)

The fanning factor can be determined from the Moody's diagram if the relative roughness and the Reynold's number of the liquid under flow is known. Moody's diagram shows the relationship between Reynold's number and absolute roughness of the pipe through which the liquid is flowing. The rougher the pipe the more turbulent the flow through that pipe. The relative roughness of a pipe is given by,

Relative roughness = ***e*/D**

where,

e = Absolute roughness (m)

D = Diameter of pipe (m)

The Reynold's number (Re) is a dimensionless parameter for characterizing fluid flow. It is the criterion used to determine whether flow is Laminar, Critical (transitional) or Turbulent. Usually for laminar flow Re < 2100, for critical flow Re is between 2100 and 4000, while turbulent flow prevails for Re > 4000.

$$\boxed{\mathbf{Re = V_p D / \upsilon = \rho V_p D / \eta}} \qquad \text{... (Eq. 10.2)}$$

where,

Re = Reynold's number

D = Diameter of the pipe (m)

V_p = Velocity at the pipe (m/s)

υ = Kinematic viscosity

ρ = Density of water (kg/m^3)

η = Viscosity of water (poise)

By noting down the values of the Reynold's number and the relative roughness the fanning factor is determined.

$$\boxed{f_f = f/4} \qquad \text{... (Eq. 10.3)}$$

where,

f_f = Fanning friction factor

f = Darcy's friction factor (obtained from the Moody's diagram)

PROCEDURE

1. Arrange the pipe straight either attached to the pump or to the tap.
2. Find out the length and internal diameter of the pipe.
3. The other end of the pipe is placed into the measuring tank.
4. Allow the liquid to flow through the pipe.
5. Find out the liquid level collected in the measuring cylinder for a pre determined time.
6. Determine the discharge (Q) of the liquid in m^3/sec.
7. Find out the velocity of discharge (m^2/sec).

8. Find out the density as well as the viscosity of the liquid flowing (usually water).
9. Calculate the Reynold's number.
10. Calculate the relative roughness of the pipe (Refer to appendix XVII).
11. Find out the Darcy's fanning factor from the Moody's diagram (Refer to appendix XVI). Then find out the Fanning Friction Factor. Substitute in the equation to determine the pressure drop at that particular velocity.
12. Repeat the experiment with different flow rate and Reynold's number and find out the pressure loss.
13. Draw a graph by taking flow rate on X axis and pressure loss on Y axis.

OBSERVATIONS AND CALCULATIONS

Length of the pipe **(L)** = ... m

Internal diameter of the pipe **(d)** = ... m

Radius of the pipe **(R_h)** = ... m

Area of the pipe **(Πr^2)** = ... m^2

Actual Discharge **(Q)** = A X R /*t*

Area of the measuring cylinder **(A)** = ...m^2

Rise in water level **(R)** = ...*x*......m

Time taken to collect *x m* of water in

The measuring cylinder = ...sec

Velocity of the discharge **(V)** = Q/ Area of the pipe

= .... m/sec

Density of water **(ρ)** = ... (Kg/m^3)

Viscosity of water **(η)** = ... (Poise)

Reynold's number **(Re)** = ρ V D /η

= ....

Absolute roughness of the pipe = ...

(Refer to appendix XVII)

Relative roughness of the pipe = *e*/D

= ....

Darcy's frictional factor **(*f*)** = ...

Fanning's factor **(f_f)** = *f*/4

= ....

Frictional pressure loss of liquid head in (m) **(h_f)** = f_f **(L/R_h) × (V^2 /2g)**

Acceleration due to gravity **(*g*)** = ...

(h_f) = ... m

Table 10.1: Determination of Frictional Pressure Loss

Sl No	*Actual discharge*	*Velocity of discharge (m/sec)*	*Reynold's number*	*Darcy's frictional factor*	*Fanning friction factor*	*Frictional pressure loss (m)*
1						
2						
3						
4						
5						
6						
7						
8						

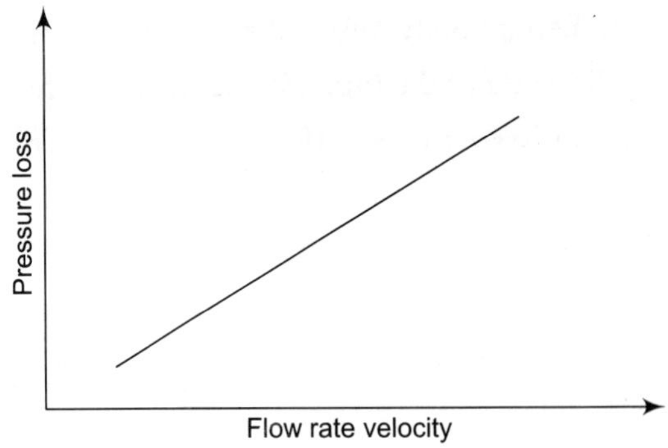

Fig. 10.1: Model graph—Effect of flow rate velocity on pressure loss.

REPORT

The pressure loss due to friction is determined during the flow of liquid through a pipe. The effect of velocity of flow on frictional loss is studied. And it is found that the as the velocity of flow/ Reynold's number .... the pressure loss is ....

EXPERIMENT 11

DETERMINATION OF CO-EFFICIENT OF DISCHARGE BY ORIFICE METER

AIM

To demonstrate the use of orifice meter as flow meter.

To determine the co-efficient of discharge.

To study the effect of flow rate on Reynolds number of the liquid.

REQUIREMENTS

Orifice Meter Assemble	(1)
Stopwatch	(1)
Water supply	
Drain	

PRINCIPLE

An orifice meter consists of a flat circular plate with circular hole called orifice which is concentric with the pipe axis. The orifice meter is fitted with the pipe line. If a constriction is placed in a closed channel carrying a stream of fluid there will be increase in the velocity and therefore increase in kinetic energy. According to the Bernoulli's theorem the total energy at any given point during flow is a constant. So because of the increment in the kinetic energy the pressure energy at the constriction is reduced to balance the increment in the kinetic energy. Rate of discharge from the constriction can be calculated by knowing this pressure reduction, the area at the constriction, the density of the fluid, and the coefficient of discharge.

The co-efficient of discharge can be determined by the following equation,

$$C_d = Q_a/Q_t$$

C_d = Co-efficient of discharge

Q_a = Actual discharge

Q_t = Theoretical discharge

$$Q_t = \frac{a_1 \cdot a_2 \, x \sqrt{2gH}}{\sqrt{a_1^2 - a_2^2}} \qquad \text{... (Eq. 11.1)}$$

where,

a_1 = Area at the inlet of orifice meter in mm^2

a_2 = Area of orifice plate in mm^2

H = 12.6 h

h = Pressure difference in m of Hg

g = Acceleration due to gravity in m/sec^2

And

$$Q_a = A \times R/t \qquad \text{... (Eq. 11.2)}$$

where,

A = Area of the measuring tank (m^2)

R = Rise of water level in measuring tank (m)

t = Time taken for rise of water level in measuring tank (sec)

$$Re_D = V_p D/\upsilon = \rho V_p D/\eta \qquad \text{... (Eq. 11.3)}$$

$$Re_d = V_o d/\upsilon = \rho V_o d/\eta$$

where,

Re_D = Reynold's number based on D

D = Diameter of the pipe

Re_d = Reynold's number based on d

d = Diameter of the orifice

V_p = Velocity of flow at the pipe (m/s)

V_o = Velocity of flow at the orifice (m/s) (Q_a /area of the orifice)

υ = Kinematic viscosity

ρ = Density of water (kg/m^3)

η = Viscosity of water (poise)

PROCEDURE

Figure shows the assemble for the determination of flow rate by orifice meter.

1. Clean the apparatus and tank properly to avoid any interruption in the experiment.
2. Check the power supply.
3. Close the water outlet valve.
4. Fill water in the tank by ¾th volume.
5. Close all the valves except by pass valve.
6. Check the level of the mercury in the manometer tube.
7. Close all pressure tapes of manometer connected to the orifice meter (P_1 & P_2).
8. Switch on the power supply and the pump.

9. Operate the flow control valve to allow the flow through the orifice meter.
10. Open the pressure taps of the manometer slowly. (care should be taken to avoid the flow of liquid to the manometer).
11. Open the air release valve to release the air.
12. When there is no air in the manometer close the air release valve.
13. Note down the manometer reading.
14. Arrange a measuring jar at the discharge end to determine the fluid discharged through the orifice meter.
15. Using a stopwatch the time required for liquid discharged can be determined.
16. Repeat the steps 10-16 for different flow rate of water by adjusting control valve.
17. Finally when all the readings are recorded open the Bypass valve fully. Then close the flow control valve.
18. Close all the manometer pressure taps.
19. Now switch of the pump and the power supply.
20. The values are tabulated and determine the coefficient of discharge at different flow rates.
21. Find of the Reynold's number for different flow rate and study the effect of flow rate on Reynold's number.

Note: A simple apparatus without pump is also available in the market. It can be directly connected pipe through which the fluid flows. The readings can be taken with different flow rate by controlling flow.

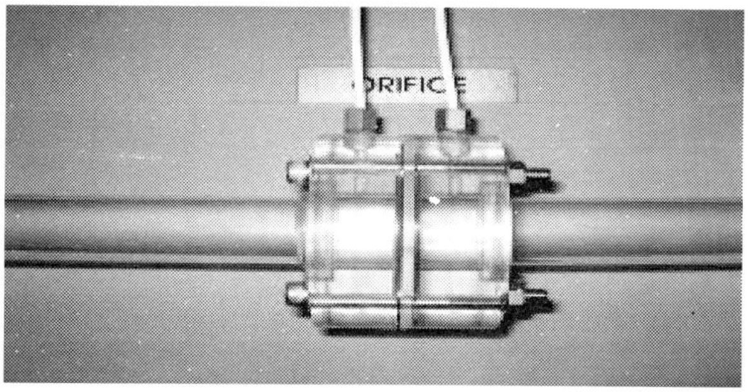

Fig. 11.1: Orifice meter.

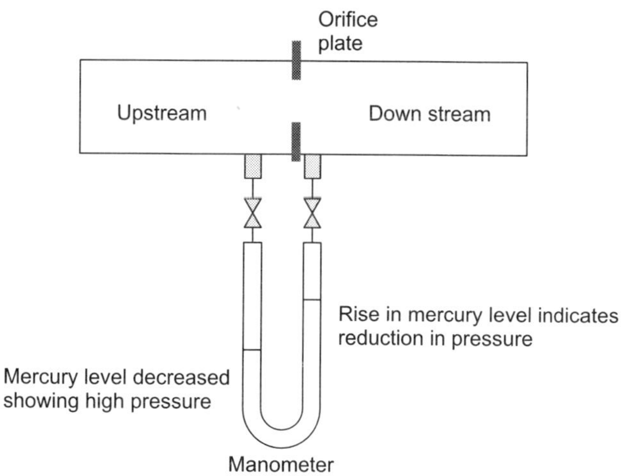

Fig. 11.2: Working of orifice meter.

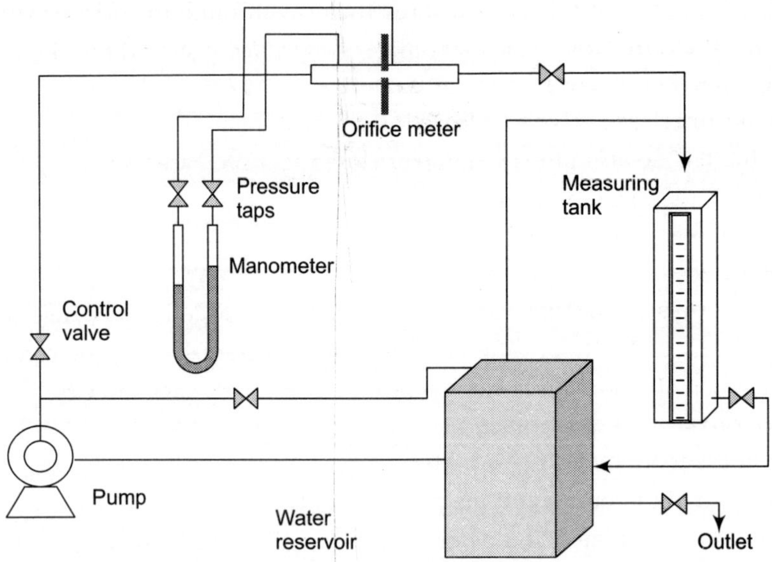

Fig. 11.3: Assemble of orifice meter for the determination of rate of flow.

OBSERVATIONS AND CALCULATIONS

Table 11.1: Determination of Pressure Difference by Varying the Flow

Sl No	h_1 *(cm)*	[illegible] *(cm)*	*t (sec)*
1			
2			
3			
4			
5			

h_1 = pressure difference (cm of Hg)

R_1 = Rise of water level in measuring tank in (Cm)

Area of the measuring tank (**A**) = ...m^2

Specific gravity of Hg (**S**) = 13.6

Acceleration due to gravity (g) = 9.81 m/ sec^2

Diameter at the inlet of the orifice meter (d_1) = ...mm

= ....m

Diameter of orifice plate (d_2) = ...mm

= ....m

Area at the inlet of orifice (a_1) = Πr^2

(r is the radius of the inlet of the orifice plate)

= .... m^2

Area of the orifice plate = Πr^2

(r is the radius of the orifice plate)

$= \ldots\ldots\ldots\ldots \text{ m}^2$

Theoretical discharge (Q_t) $= a_1 . a_2 \text{ X } \sqrt{2gH} / \sqrt{a_1^2 . a_2^2}$

Actual discharge (Q_a) $= \dfrac{AXR}{t}$

Co-efficient of discharge (C_d) $= Q_a / Q_t$

where,

$H = 12.6 \text{ X } h$

h = Pressure difference in m of Hg

A = Area of measuring tank (m^2)

R = Rise of water level in the measuring tank (m)

t = Time taken for rise of water level in measuring tank (sec)

$V_p = Q_a$/area of the pipe

r_p (radius of the pipe) = ...cm

= .... m

Area of the pipe $= \Pi r_p^2$

$= \ldots\ldots\ldots \text{ m}^2$

Radius of the orifice (r_o) = ... cm

= ....m

Area of the orifice $= \Pi r_o^2$

Velocity of flow with respect pipe (V_p) $= Q_a / \Pi r_p^2$

= ....m/s

Velocity of flow with respect orifice (V_o) $= Q_a / \Pi r_o^2$

Viscosity of the water (η) = ... Poise

Density of water (ρ) = ... g/cc

$= \ldots\ldots\ldots \text{Kg/m}^3$

$Re_D = \rho V_p D / \eta$

= ....

$Re_d = \rho V_o d / \eta$

= ....

Table 11.2: Determination of Co-efficient of Discharge

Sl No	*h(m)*	*R(m)*	*H(m)*	Q_a *m³/s*	Q_t *(m³/sec)*	$C_d = Q_a/Q_t$
1						
2						
3						
4						
5						

Table 11.3: Determination of the Effect of Flow Rate on Reynold's Number

Sl No	Q_a (m^3/sec)	V_p (m/s)	V_o (m/s)	Re_D	Re_d
1					
2					
3					
4					
5					

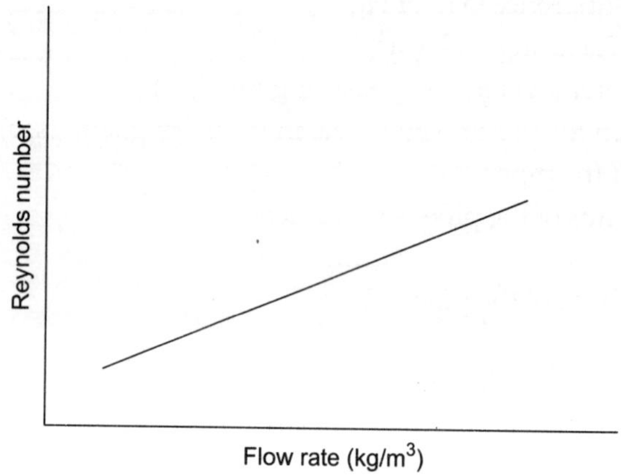

Fig. 11.4: Model graph—Effect of flow rate on Reynold's number.

REPORT

The average discharge of the orifice meter is = ……….. m^3/sec
The average Co-efficient of discharge C_d = ……….
As the flow rate increase the Reynolds number ……………..

EXPERIMENT 12

DETERMINATION OF CO-EFFICIENT OF DISCHARGE BY VENTURI METER

AIM

To demonstrate the use of venturi meter.

To determine the co-efficient of discharge during fluid flow.

To study the effect of flow rate on Reynold's number of liquid.

REQUIREMENTS

Venturi meter assemble	(1)
Stopwatch	(1)
Water supply	
Drain	

PRINCIPLE

To control the industrial process, it is essential to know the material entering and leaving the process. If materials are transported in the form of fluids, the determination of flow rate is so important. Venturi meter is one of the flow meters commonly used in the industry for the determination of flow rate of fluids.

A venturi meter consist of a short conical inlet section which leads to throat section then to a long discharge cone. Manometer connections are done at the start of the inlet section and at the throat of the venturi meter. When the fluid is entering to the constricted portion (throat of venturi) there will be an increase in the velocity, and hence increase in the kinetic energy. So according to the Bernoulli's theorem there should be a reduction in pressure energy. This pressure drop in the cone is used to determine the flow rate. It is also desirable to know the density of the fluid flowing, diameter ratio, and venturi coefficient of discharge. The coefficient of discharge is determined experimentally. For a well designed venturi meter this will be about 0.98 for pipe diameter of 2-8 inches and 0.99 for large sizes.

The co-efficient of discharge can be determined by the following equation,

$$\boxed{C_d = Q_a/Q_t} \quad \text{... (Eq. 12.1)}$$

C_d = Co-efficient of discharge
Q_a = Actual discharge
Q_t = Theoretical discharge

$$\boxed{Q_t = \frac{a_1 \cdot a_2 \times \sqrt{2gH}}{\sqrt{a^2{}_1 - a^2{}_2}}} \quad \text{... (Eq. 12.2)}$$

where,

a_1 = Area at the inlet of venturi meter in m^2
a_2 = The area at throat of the venturi meter in m^2
H = 12.6 h
h = Pressure difference in m of Hg
g = Acceleration due to gravity in m/sec^2

And,

$$\boxed{Q_a = A \times R/t} \quad \text{... (Eq. 12.3)}$$

where,

A = Area of the measuring tank (m^2)
R = Rise of water level in measuring tank (cm)
t = Time taken for rise of water level in measuring tank (sec)

A discharge co-efficient $c_d = 0.60$ may be taken as standard, but the value varies noticeably at low values of the Reynolds number.

$$\boxed{\begin{aligned} Re_D &= V_p\, D/\upsilon = \rho\, Vp\, D/\eta \\ Re_d &= V_o\, d/\upsilon = \rho\, V_o\, d/\eta \end{aligned}} \quad \text{... (Eq. 12.4)}$$

where,

Re_D = Reynold's number based on D
D = Diameter of the pipe (m)
Re_d = Reynold's number based on d
d = Diameter of the throat of venturi (m)
V_p = Velocity at the pipe m/s
V_o = Velocity of flow at the throat of venturi (m/s) (Q_a/area of the throat of venturi)
υ = Kinematic viscosity
ρ = Density of water (kg/m^3)
η = Viscosity of water (poise)

Fig. 12.1: Venturi meter.

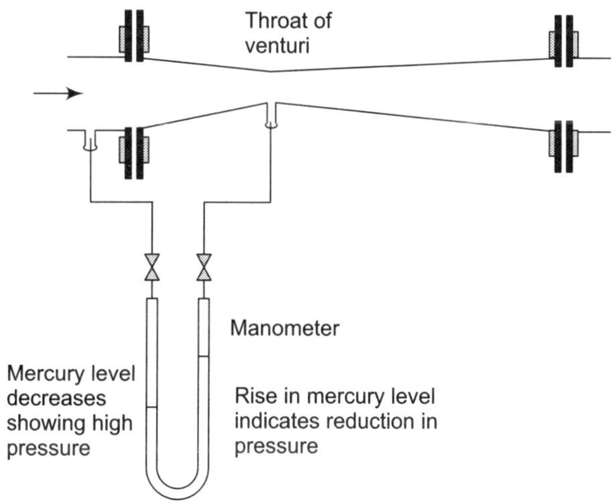

Fig. 12.2: Working of venturi meter.

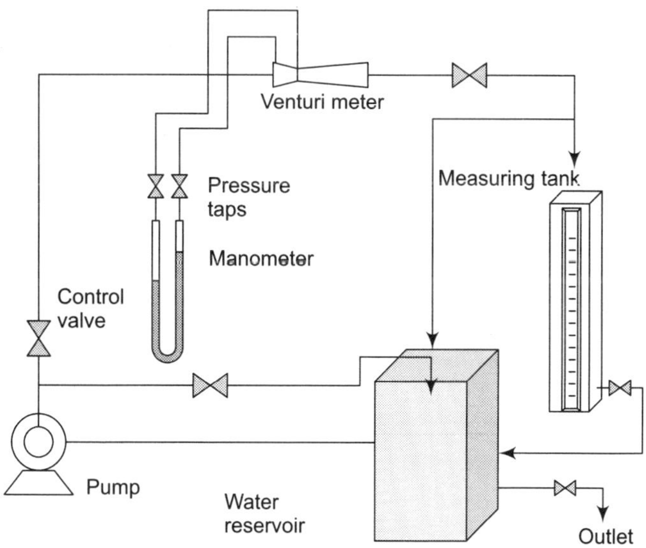

Fig. 12.3: Assemble of venturi meter for the determination of rate of flow.

PROCEDURE

Figure 12.3 shows the assemble of venturi meter.

1. Clean the apparatus and tank properly to avoid any interruption in the experiment.
2. Check the power supply.
3. Close the outlet (drain valve).
4. Fill water in the tank by ¾th of the total volume.
5. Close all the valves except by pass valve.
6. Check the level of mercury in manometer tube.
7. Close all pressure tapes of manometer connected to the orifice meter (P_1 & P_2).
8. Switch on the power supply and the pump.
9. Operate the flow control valve to allow the flow through the orifice meter.
10. Open the pressure taps of the manometer slowly, care should be taken to avoid the flow of liquid to the manometer.
11. Open the air release valve to release the air.
12. When there is no air in the manometer close the air release valve.
13. Note down the manometer reading.
14. Arrange a measuring jar at the discharge end to determine the fluid discharged through the orifice meter.
15. Using a stopwatch the time required for liquid discharged can be determined.
16. Repeat the steps 10-16 for different flow rate of water by adjusting control valve.
17. Finally when all the readings are recorded open the Bypass valve fully. Then close the flow control valve.
18. Close all the manometer pressure taps.
19. Now switch of the pump and the power supply.
20. The values are tabulated and determine the coefficient of discharge.
21. A graph is drawn by taking flow rate on X axis and Reynold's number on Y axis to understand the effect of velocity of flow (V_p or V_o) on Reynold's number (Fig. 12.4).
22. Another graph is drawn by taking co-efficient of discharge on X axis and Reynold's number on Y axis. (Fig. 12.4).

Note: A simple apparatus without pump is also available in the market. It can be directly connected pipe through which the fluid flows. The readings can be taken with different flow rate by controlling the flow.

OBSERVATIONS AND CALCULATIONS

Table 12.1: Determination of Pressure Difference by Varying the Flow

Sl No	h_1 *(cm)*	R_1 *(cm)*	*t (sec)*
1			
2			
3			

h_1 = Pressure difference (cm of Hg)

R_1 = Rise of water level in measuring tank (cm)

Area of the measuring tank (**A**) = ... m^2

Specific gravity of Hg (**S**) = 13.6

Acceleration due to gravity (g) = 9.81 m/ sec^2

Diameter at the inlet of the venturi meter (d_1) = ... mm

= .... m

Diameter of throat of venturi (d_2) = ... mm

= .... m

Area at the inlet of orifice meter (a_1) = Πr^2

(r is the radius of the inlet of the venturi meter)

= .... m^2

Area of the throat of venturi = Πr^2

(r is the radius of the throat of venturi)

= .... m^2

$$\text{Theoretical discharge } (\mathbf{Q}_t) = \frac{a_1 \cdot a_2 \times \sqrt{2gH}}{\sqrt{a^2{}_1 - a^2{}_2}}$$

$$\text{Actual discharge } (\mathbf{Q}_a) = \frac{\text{AXR}}{t}$$

Co-efficient of discharge ($\mathbf{C}_d$) = Q_a/Q_t

where,

H = 12.6 × h

h = Pressure difference in m of Hg

A = Area of measuring tank (m^2)

R = Rise of water level in the measuring tank (m)

t = Time taken for rise of eater level in measuring tank (sec)

V_p = Q_a/Area of the pipe

r_p (radius of the pipe) = ... cm

= .... m

Area of the pipe = Πr_p^2

= .... m^2

Radius of the throat of venturi (r_o) = ... cm

= .... m

Area of the throat of venturi = Πr_o^2

Velocity of flow with respect pipe ($\mathbf{V}_p$) = $Q_a/\Pi r_p^2$

= .... m/s

Velocity of flow with respect to throat of venturi ($\mathbf{V}_o$) = $Q_a/\Pi r_o^2$

Viscosity of the liquid (water) (η) = ... Poise

Density of water (ρ) = ... g/cc

= .... Kg/m^3

$Re_D = \rho V_p D/\eta$

= ....

$Re_d = \rho V_o d/\eta$

= ....

Table 12.2: Determination of Co-efficient of Discharge

Sl No	h (m)	R (m)	H (m)	Q_a m^3/s	Q_t (m^3/sec)	$C_d = Q_a/Q_t$
1						
2						
3						

Table 12.3: Effect of Flow Rate on Reynold's Number

Sl No	Q_a (m^3/sec)	V_p (m/s)	V_o (m/s)	Re_D	Re_d
1					
2					
3					
4					
5					

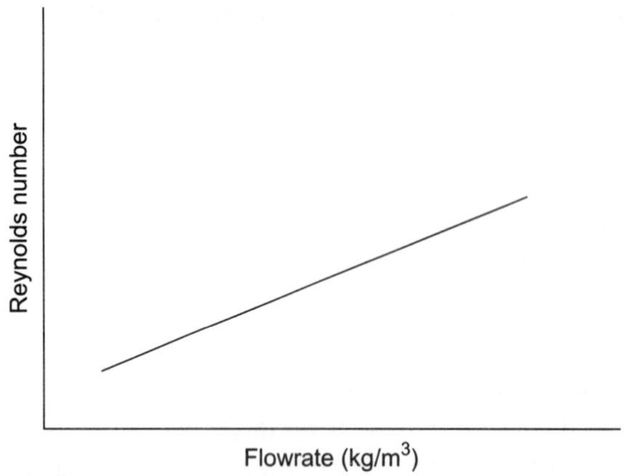

Fig. 12.4: Effect of flow rate on Reynold's Number.

REPORT

The average discharge of the venturi meter is = ... m^3/sec

Co-efficient of discharge C_d = ...

As the flow rate increase the Reynolds number ....

EXPERIMENT 13

DETERMINATION OF EFFICIENCY OF A CENTRIFUGAL PUMP

AIM

To determine the co-efficient of discharge of a centrifugal pump.

To determine the overall efficiency of centrifugal pump.

To determine the efficiency of the centrifugal pump.

REQUIREMENTS

The centrifugal pump assemble	(1)
Stopwatch	(1)
Energy meter	(1)

PRINCIPLE

Pump is a device which converts mechanical energy to hydraulic energy. Hydraulic energy is in the form of Pressure energy. In a centrifugal pump the mechanical energy is converted into pressure energy by centrifugal force.

A centrifugal pump is a roto dynamic pump that uses a rotating impeller to increase the pressure of a fluid. They are commonly used to move liquids through a piping system. Its purpose is to convert energy of a prime mover (an electric motor or turbine) first into velocity or kinetic energy and then into pressure energy of a fluid that is being Pumped. The energy change occurs by virtue of two main parts of the pump, the impeller and the volute or diffuser. The impeller is the rotating part that converts driver energy into the kinetic energy. The volute or diffuser is the stationary part that converts the kinetic energy into pressure energy. The overall system efficiency is dependent on motor efficiency, pump efficiency (velocity of the impeller, impeller diameter, number of blades etc) and the total head developed.

$$\text{The total efficiency of the system} = \frac{\text{Output}}{\text{Input}} \times 100 \quad \text{... (Eq. 13.1)}$$

The output of the pump system can be determined by,

$$\text{Output} = \frac{Q \times H \times g \times \rho}{1000}\ \text{KW} \quad \text{... (Eq. 13.2)}$$

where,

Q = Flow rate (Discharge) in m^3/sec
H = Total head in m
g = Acceleration due to gravity m/s^2
ρ = Density of the liquid in kg/m^3

The motor input power can be measured by using portable power analyzers. When we discuss centrifugal pumps we can compare the amount of work or power we get out of the pump to the amount of power we are putting into the pump.

Remember that we are using the actual horsepower or kilowatts going into the pump and not the horsepower or kilowatts required for the electric motor. Most motors run somewhere near 85% efficiently. So the efficiency of the pump can be determined from the equation,

$$\text{Efficiency of the pump} = \text{Total efficiency}/ .85 \quad \text{... (Eq. 13.3)}$$

A survey of popular pump brands demonstrates that pump efficiencies range from 15% to over 90%.

Description of Some Important Parameters

All of the forms of energy involved in a liquid flow system are expressed in terms of feet of liquid *i.e.*, head.

Suction Head (Suction Lift) h_s

It is the vertical height of the central line of the centrifugal pump above the water surface in the tank or pump from which water is to be lifted.

Delivery Head (h_d)

The vertical distance between the central line of the pump and water surface in the tank to which water is to be delivered is known as delivery head.

Total Head (H)

Total head is

1. The work of the pump.
2. The measure of the pump's ability to raise the liquid to a given height.
3. The measure of the pump's ability to develop a given discharge pressure.
4. The discharge elevation minus the suction elevation.
5. The discharge head minus the suction head.
6. The discharge head plus the suction lift.
7. The discharge absolute pressure reading minus the suction absolute pressure reading.

$$\text{Total head} = \text{delivery head} + \text{suction head} \quad \text{... (Eq. 13.4)}$$

Suction Pressure

Suction pressure is the pressure at the pump's suction nozzle as measured on a gauge. The suction pressure is probably the most important pressure inside the pump. The entire pump's production is based on the suction pressure. The pump takes suction pressure and converts it into discharge pressure. If the suction pressure is inadequate, it leads to cavitation. Because of this, all pumps need a gauge at the suction nozzle to measure the pressure entering the pump.

Discharge Pressure

This is taken from the pump discharge side pressure gauge. Installation of the pressure gauge in the discharge side is a must, if not already available.

Pump Capacity, Q = Volume of liquid delivered by pump per unit time, m^3/hr or m^3/sec

Q is proportional to N, where N rotational speed of the pump.

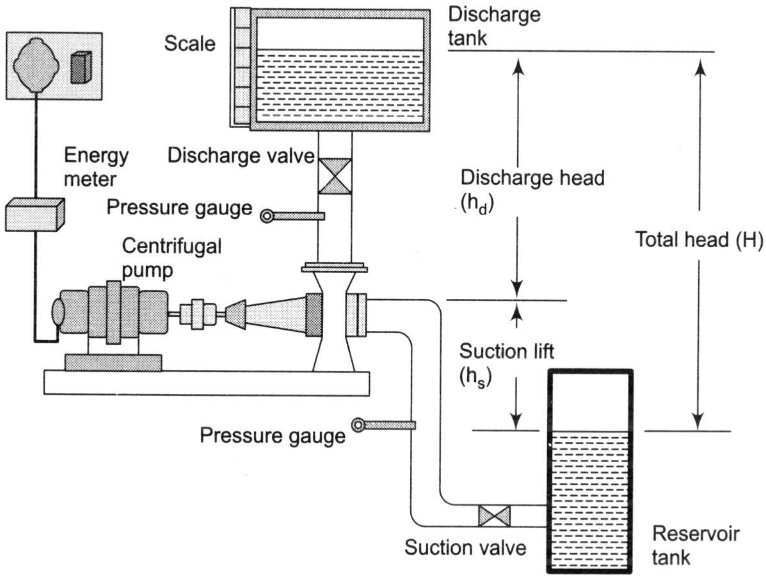

Fig. 13.1: Centrifugal Pump assemble.

Energy meter

The energy meter is an electrical measuring device, which is used to record Electrical Energy Consumed over a specified period of time in terms of units. It consist of a flat metal plate that is circular and rotates. There is a black mark on the plate so that you can count the number of the rotations. The rate of rotation, usually measured in rotations per minute (r/min.), relates directly to the power being used at the time.

PROCEDURE

1. Clean the reservoir tank as well as the collection tank.
2. Fill the water in the reservoir tank.

3. Check the assembly.
4. Vent the pump and close it.
5. Check the pressure gauzes and the energy meter.
6. Find out the area of the collection tank.
7. Open the suction valve and discharge valve.
8. Switch on the power.
9. Start the stopwatch.
10. Find out the time required for 5 revolutions of the disc in the energy meter.
11. Simultaneously find out the time required for filling 20 cm length in the collection tank.
12. Note down the readings.
13. Note down the pressure of suction from the pressure gauze (I) at the suction end (p_s).
14. Pressure of discharge (p_d) is noted down from the pressure gauze (II) at the discharge end.
15. The total head is calculated.
16. The steps 7-13 are repeated twice to avoid errors.
17. Switch off the power.
18. Water from the collecting tank is released to the reservoir tank.
19. With the help of equations find out the input, output, co-efficient of discharge, and the overall efficiency of the pump.

Precautions

- Do not pump if the supply tank is dry.
- Be sure that the reservoir tank is clear and will not freeze in cold weather.
- Remember that constant running is easier on bearings and seals.

OBSERVATIONS AND CALCULATIONS

Suction pressure (p_S) = ...

= ......*a*..... mmHg

Discharge pressure (p_d) = ...

= ......*b*....mmHg

Area of the collection tank (A) = ... m^2

Acceleration due to gravity (g) = 9.81 m/s^2

Density of water = 996 kg/m^3

Motor efficiency = 85%

Table 13.1: Determination of Total Head (H)

Sl No	Suction pressure (p_s) mmHg	Discharge pressure (p_d) mmHg	Suction head (m)	Discharge head (m)	Total head (m)
1					
2					
3					

Table 13.2: Determination of T and *t*

Sl No	*Time for 20 cm rise in the collection tank (sec) T*	*Number of revelations of the energy meter (x)*	*Time for 5 revolutions of the disc in the energy meter (t)*
1		5	
2		5	
3		5	

Table 13.3: Determination of Efficiency of the Pump

Sl No	*Total head (H)*	*Discharge (Q) m³/sec*	*Output (KW)*	*Input (KW)*	*Total efficiency (%)*	*Pump efficiency (%)*
1						
2						
3						

Determination of Total Head

(Convert the pressure reading from the gauge in to mmHg. See the Appendix I &II)

760 mm Hg = 10.33 m of water

......*a*.....mmHg = ...*c*......... m of water.

Suction head (suction lift) (h_s) = ... *c* ....m of water.

.........*b*.... mmHg = ...*d*.........m of water.

Discharge head (h_d) = ... *d*.........m of water.

Total head (**H**) = $h_s + h_d$

= ....m

$$\text{Discharge (flow rate) (Q)} = \frac{A \times H}{T}$$

= .... m³/sec

$$\text{Output} = \frac{Q \times H \times g \times \rho}{1000} \text{ KW}$$

= .... KW

Input = $x/t \times 3600 \times 1/750$ KW

(750 revolutions of the disc in the energy meter consumes 3600 units of energy)

= .... KW

Total efficiency of the system = Output /Input × 100

= ....%

Pump efficiency = Total efficiency/0.85

= ....%

REPORT

The overall efficiency of the pump system = ...%

The efficiency of the pump = ...%

EXPERIMENT 14

DETERMINATION OF EFFECT OF NUMBER OF BALLS ON GRINDING OPERATION OF BALL MILL

AIM

To study the effect of number of balls on grinding operation of a ball mill.

REQUIREMENTS

Ball mill	(1)
Balls (same size and density)	(Quantity sufficient)
Sieve 120 mesh	(1)
Weighing balance	(1)
Butter paper	(1)
Spatula	(1)
Weighing box	(1)
Sugar	

PRINCIPLE

Ball mill consist of a horizontally rotating hollow cylinder with length slightly greater than its diameter. The mill is partially filled with balls of steel or pebbles. In ball mill size reduction happens by impact and attrition.

The speed of rotation of the mill is important, as the slow speed of rotation contribute only a sliding action and high speed a centrifugal action. At both this speed, no significant size reduction happens. The ideal speed of a ball mill is a speed at which the balls just begins to centrifuge within the mill contributing a cascading action and hence a significant size reduction. The critical speed of rotation can be determined by the following equation.

$$\boxed{C_S = 76.63/\sqrt{D}} \quad \text{... (Eq. 14.1)}$$

where,

C_S = Critical speed in revolutions per min

D = Inside diameter of the ball mill in feet

At and above the critical speed no significant size reduction occurs. So the ideal speed for the size reduction is considered as 60-85% of the critical speed. The slower speed *i.e.*, 60-70% of the critical speed is meant for fine grinding. The optimum speed for a normal operation can be determined by,

$$\boxed{N = 57 - 40 \log D} \quad \text{... (Eq. 14.2)}$$

where,

N = Optimum speed

D = Inside diameter of the mill in feet

Other factors affecting the size reduction are the size of the ball, density of the ball, extent of operation, the amount of material as feed etc. Increasing the total weight of the balls of a given size increases the fineness of the product. The weight of the ball charge can be increased by increasing the number of balls or by using balls composed of materials with higher density.

In this experiment we are observing the effect of number of balls (of same density) on size reduction by increasing the number of balls in each operation.

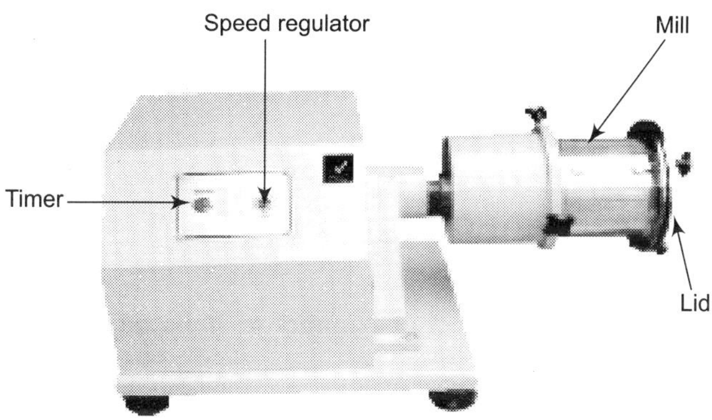

Fig. 14.1: Ball Mill.

PROCEDURE

1. Clean the ball mill thoroughly.
2. Note down the inner diameter of the mill.
3. Convert the reading in to inch. (1 inch = 2.54 cm)
4. Find out the critical speed and optimum speed of rotation.
5. Take approx 600 gm sugar and sieve it with the help of 120 mesh sieve. Remove all the fines.
6. Make 5 packets of sugar (Each containing 100 gms)
7. 100 gm of sugar (1st packet) is added to the mill.
8. Charge the mill with 8 balls of same size and density.
9. Close the lid of the mill tightly.

10. Set the time (10 min) in the timer.
11. The mill is allowed to rotate at its optimum speed for 10 min.
12. After the completion of 10 min stop the mill.
13. Take out all the contents from the mill.
14. Remove the balls from the sugar powder.
15. Transfer the sugar powder in to a 120 sieve mesh.
16. Sieve the sugar by 120 sieve mesh.
17. Collect the fines.
18. Weigh accurately the fines.
19. Find out the percentage of fines.
20. Now the second packet (100 gm of sugar) is emptied to the mill.
21. Add 12 balls of same size and density as that of the first operation to the mill.
22. Close the lid of the mill tightly.
23. Set the time (10 min) in the timer.
24. Remove the contents from the mill after the completion of the 10 min.
25. Separate the balls and the sugar is sieved by 120 meshes.
26. Find out the weight of the fines.
27. Determine the % of fines.
28. The steps 20 – 27 is repeated for next 3 packets of sugar with 16, 20, and 24 number of balls.
29. The observations are recorded.
30. Plot a graph by taking number of balls in X axis and % of fines in Y axis.

OBSERVATIONS AND CALCULATIONS

Table 14.1: Determination of Effect of Number of Balls on Grinding Operation

Sl No	*Initial weight of the powder (gm)*	*Number of balls*	*Weight of fines (gm)*	% weight of fines
1	100	8		
2	100	12		
3	100	16		
4	100	20		
5	100	24		

Critical speed of rotation of the ball mill

$$C_S = 76.63/\sqrt{D}$$

where,

C_S = Critical speed of rotation in rpm

D = Inner diameter of the ball mill in inch

D = ...cm

= ....inch

C_S = ...rpm

Optimum speed

$$N = 57 - 40 \log D$$

N = Optimum speed

D = Inside diameter of the mill in feet

N = ...

$$\% \text{ of fines} = \frac{\text{Weight of fines}}{\text{Initial weight}} \times 100$$

= ....

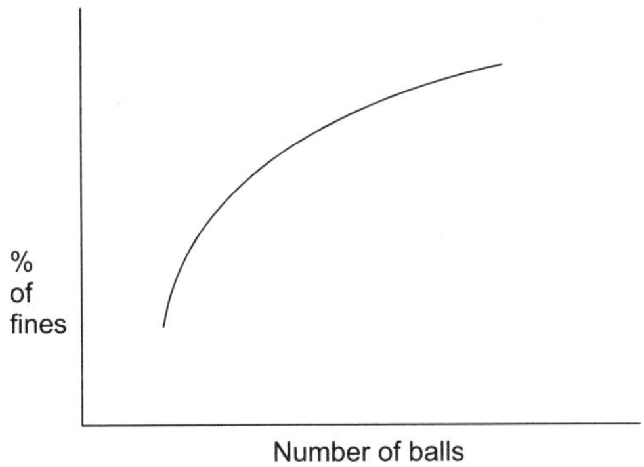

Fig. 14.2: Model graph.

REPORT

Percentage of fines increases with increase in the number of balls.

Hence size reduction is increased by increasing the number of balls.

The critical speed of rotation of the ball mill = ...

The optimum speed of rotation of ball mill = ...

EXPERIMENT 15

DETERMINATION OF EFFECT OF TIME ON GRINDING OPERATION OF A BALL MILL

AIM

To study the effect of time on grinding rate of a ball mill.

REQUIREMENTS

Ball mill	(1)
Balls (same size and density)	(12)
Sieve 120 or 85 mesh	(1)
Weighing balance	(1)
Butter paper	(1)
Spatula	(1)
Weighing box	(1)
Sugar	

PRINCIPLE

Ball mill consist of a horizontally rotating hollow cylinder shape with the length slightly greater than its diameter. The mill is partially filled with balls of steel or pebbles. In ball mill size reduction happens by impact and attrition. The speed of rotation is so important, as the slow speed of rotation contribute only a sliding action and high speed a centrifugal action. At both this speed, no significant size reduction happens. The optimum speed of a ball mill is a speed at which the balls just begins to centrifuge within the mill contributing a cascading action and hence a significant size reduction.

The optimum speed of rotation can be determined by the following equation,

$$\boxed{N = 57 - 40 \log D} \quad \text{...(Eq. 15.1)}$$

where,

N = Optimum speed in rotation per min

D = Inside diameter of the ball mill in feet

Other factors affecting the size reduction are the size of the balls, density of the balls, and extent of operation, the amount of material as feed etc. if the duration of the operation is more the amount of size reduction achieved will be more. In this experiment the materials are subjected to size reduction for different extent of time and the amount of size reduction achieved is determined.

PROCEDURE

1. Clean the ball mill thoroughly.
2. Note down the inner diameter of the mill.
3. Convert the reading into inch. (1 inch = 2.54 cm)
4. Find out the optimum speed of rotation.
5. Take approximately 500 gm sugar and sieve it with the help of 120 mesh sieve. And remove all the fines.
6. Add this sugar into the mill.
7. Add 12 balls of same size and density.
8. Close the lid tightly.
9. Set the timer for 5 min and operate the mill at its optimum speed.
10. After 5 min take out the contents from the mill.
11. Pass the content through 120 mesh and collect the fines.
12. Weigh the fines obtained.
13. Weigh the rest of the sugar and note the weight.
14. Add the sugar into the mill again.
15. Operate it for another five minutes.
16. After the time of 5 min take out the contents.
17. Pass the sugar through sieve number 120 mesh.
18. Weigh the fines obtained.
19. This process is continued for another 15, 30, 45 min.
20. The values are tabulated.
21. Find out the % of fines obtained after each operation.
22. A graph is plotted by taking time on X axis and percentage of fines on Y axis.

OBSERVATIONS AND CALCULATIONS

Table 15.1: Determination of Effect of Time on Grinding

Sl No	*Time (min)*	*Initial weight (gm)*	*Weight of fines (gm)*	*Cumulative weight of fines*	*% of fines*
1	5		(x)	(x)	

2	10		(y)	$(x+y)$	
3	15		(z)	$(x+y+z)$	
4	30				
5	45				
6	60				

$$\% \text{ of fines} = \frac{\text{Cumulative weight of fines}}{\text{Initial weight of sugar}} \times 100$$

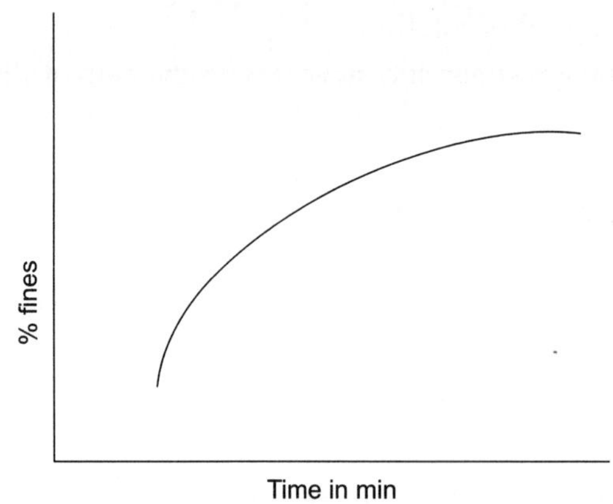

Fig. 15.1: Model graph.

REPORT

Effect of time on grinding operation of ball is studied and found that as the time of operation increases the % of fines ....

EXPERIMENT 16

SIZE REDUCTION – HOMOGENIZATION

AIM

To reduce the size of the oil globules in an emulsion.

REQUIREMENTS

Silverson homogenizer	(1)	Castor oil
Beakers	(2)	Acacia
Mortar and pestle	(1)	Water
Glass rod	(1)	
Test tube	(2)	
Test tubes stand	(1)	
Measuring cylinder	(1)	
Stage micrometer	(1)	
Eye piece micrometer	(1)	
Microscope	(1)	
Slide	(2)	
Cover slip	(2)	
Blotting paper		

PRINCIPLE

Emulsions are defined as the biphasic liquid preparations containing two immiscible liquids, (oil and water) in which one liquid is dispersed as small globules in the other with the help of an emulsification agent by applying mechanical force. In conventional emulsification process using mortar and pestle, the mechanical energy given is comparatively less so that the dispersed globules are comparatively larger and not uniform in size. This will leads to the instability of emulsion while storage.

Homogenization is the process of converting two immiscible bulk phase liquids into an emulsion (primary homogenization) or of reducing the size of the droplets in a pre-existing emulsion (secondary homogenization) this can be achieved by passing the fluid through restricted orifices under high pressure. *e.g.*, silverson emulsifier.

The high speed rotation of the rotor blades creates a powerful suction which draws the liquid into the work head. The materials are subjected to intense shear in the confined area of the work head. Agglomerates are broken down in the gap between the rotor blades and stator wall. A lump free mixture is rapidly obtained. The product is forced out of the stator as fresh material is drawn into the work head. A circulatory mixing cycle develops in which all the material passes through the Silverson work head produces efficient mixing and desirable size reduction.

The particle size and size distribution of the oil globules in the emulsion are determined by optical microscopy. The result obtained is compared with the one obtained before homogenization.

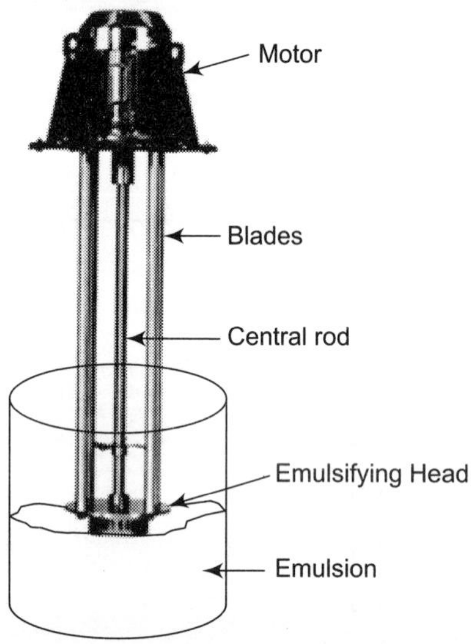

Fig. 16.1: Silverson Homogenizer.

PROCEDURE

I. Preparation of emulsion

1. Emulsion is prepared with the help of motor and pestle.
2. The standard formula for the emulsion is

   Castor oil = 37.5 ml

   Acacia = 10.9 gm

   Water to produce = 100 ml.

3. Transfer the prepared emulsion into a clean beaker.

II. Determination of average particle size and size distribution of oil globules in the emulsion

Calibration of eyepiece micrometer

1. Select the 45X eyepiece for the experiment.
2. Place carefully the eyepiece micrometer in to 45X eyepiece.
3. Place the stage micrometer on the stage of the microscope.
4. Focus the stage micrometer scale with the help of 10X.
5. Carefully turns eyepiece diaphragm and fix the 45X.
6. Focus the scale.
7. By adjusting the stage micrometer as well as the eyepiece micrometer, bring the 0 readings on the scales together. (as shown in the Figure 16.2)
8. Now find out the next coinciding lines of the scales.
9. The eyepiece micrometer is calibrated as follows,
   Number of divisions present in the stage micrometer = 100
   Length of the stage micrometer scale = 1 mm
   1 division of stage micrometer = 1/100
   = 0.01 mm (10 μm)
   Number of divisions in the eyepiece micrometer = 100
   Length of the eyepiece micrometer scale = 10 mm

........Division of eyepiece micrometer coincides with ....division of stage micrometer.

1 division of eyepiece

$$\text{micrometer is} = \frac{\text{Number of division of stage micrometer coinciding}}{\text{Number of division of eyepiece micro meter coinciding}} \times 10\ \mu m$$

= ....μm

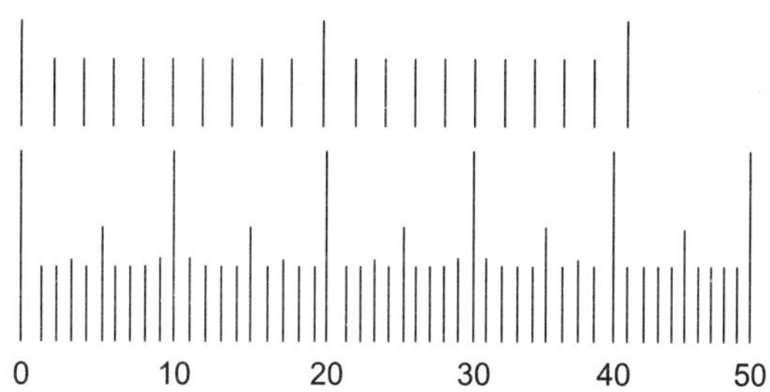

Fig. 16.2: Calibration of stage micrometer.

(Here 10th division of stage micrometer coincides with 5th division of eyepiece micrometer)

III. Determination of average size and size distribution of particles

1. Remove the stage micrometer from stage of the microscope and place it carefully in its casing.
2. Take 1 ml of the prepared emulsion on a test tube and add 10 ml of water.
3. Take one glass slide and clean it properly.
4. Put one drop of diluted emulsion in the glass slide.
5. Place the cover slip over it and wipe the sides if the excess amount flows out.
6. Place the glass slide on the stage of the microscope.
7. Focus the oil globules at 10X of the eyepiece.
8. Turn in to the 45X and focus the oil globules.

9. Bring each of the globules in between the scale divisions and count the number of divisions covered by each oil globules.
10. Minimum 100 particles are counted.
11. The values are tabulated.
12. Find out the size of each globule by multiplying the number of divisions occupied by the globule by the value obtained for 1 division of eyepiece micrometer.
13. From the values obtained find out the convenient size ranges.
14. Count the number of particles coming under each size range.
15. Tabulate the data and find out the average particle size of oil globules present in the emulsion.
16. Size distribution is shown in the form of bar diagram by taking size range in X axis and number of particles in each range on Y axis.

IV. Homogenization of emulsion by silverson emulsifier

1. Select and fit the appropriate screen, fine screens for thin liquids, coarse screens for viscous liquids.
2. Place the liquid container under the homogenizer head and lower the head into the liquid ensuring a distance of at least 1 cm is maintained between the bottom of the container and the homogenizer head.
3. Select a slow speed to begin the process and then increase the speed gradually to homogenize the liquid.
4. Homogenize for 3–5 min.
5. On completion of homogenization, slow the speed down gradually and then bring the machine to a stop before removing the liquid container for examination.
6. Take 1 ml of the homogenized emulsion in a test tube and dilute it with 10 ml of water.
7. Find out the average size and size distribution of the particles as discussed earlier.

OBSERVATIONS AND CALCULATIONS

Number of divisions occupied by oil globules

(Before homogenization)

(After homogenization)

Size of the oil globules (Number of divisions X calibrated value of eyepiece)

(Before homogenization) (After homogenization)

Table 16.1: Determination of Average Size of Oil Globules (Before Homogenization).

Sl No	*Range of particle size (μm)*	*Mean size range (μm) (d)*	*Number of particle in each range (n)*	$n \times d$
1				
2				
3				
4				
5				
6				
7				
8				
9				
10				
11				

Σn = ……….. $\Sigma n d$ = ……….

Table 16.2: Determination of Average Particle Size (After Homogenization)

Sl No	*Range of particle size (μm)*	*Mean size range (μm) (d)*	*Number of particle in each range (n)*	$n \times d$
1				
2				
3				
4				
5				
6				
7				
8				
9				
10				
11				

Σn = ……….. $\Sigma n d$ = ……….

Average particle size $= \Sigma n\,d / \Sigma n$

Average particle size before homogenization = ...µm

Average particle size after homogenization = ...µm

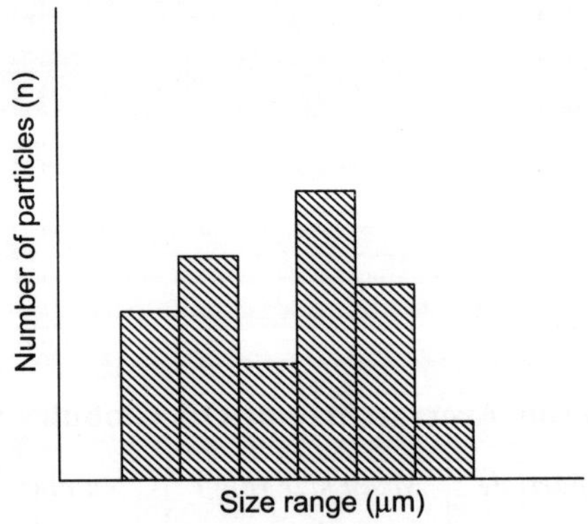

Fig. 16.3: Model graph—Size Distribution of oil globules in the emulsion.

REPORT

The average size of oil globules before homogenization =.........µm.

The average size of oil globules after homogenization =.........µm

The size distribution of oil globules is studied in the form of bar diagram.

EXPERIMENT 17

PARTICLE SIZE SEPARATION BY SIEVEING

AIM

To perform the particle size separation of the given sample and determine the arithmetic mean of the given sample.

REQUIREMENTS

Sieves of different mesh size	(Quantity sufficient)
Mechanical shaker	(1)
Weighing balance	(1)
Weighing box	(1)
Butter paper	
Powder sample	

PRINCIPLE

Size separation can be defined as a unit operation that involves the separation of various sizes of particles into two or more portions by means of screening surfaces. Performance of size reduction equipments, or crushing and grinding equipments involve the determination of the amount of material of different size present after the process. The only general and simple method for this is to determine the fraction of the sample that will go through a screen with known size of openings. A screen analysis of the sample is performed by placing a sample on the coarsest of a set of standard screens. Below this, the screens are arranged in the order of decreasing size of the openings in the mesh. The total assemble is shaken with help of a mechanical oscillator or manually, for a definite length of time and the material collected on each sieve is removed and weighed. The result obtained can be represented graphically.

1. The frequency-size distribution plot (the weight percent of the material retained Vs the average particle size).
2. Log probability curve (log of the screen aperture Vs the cumulative percentage under size or over size).
3. Cumulative size distribution curves (screen opening Vs fraction of total weight of particles having a size greater than or less than the given opening).

The mean particle size can be determined by the equation

Arithmetic mean = Weight size/100 ... (Eq. 17.1)

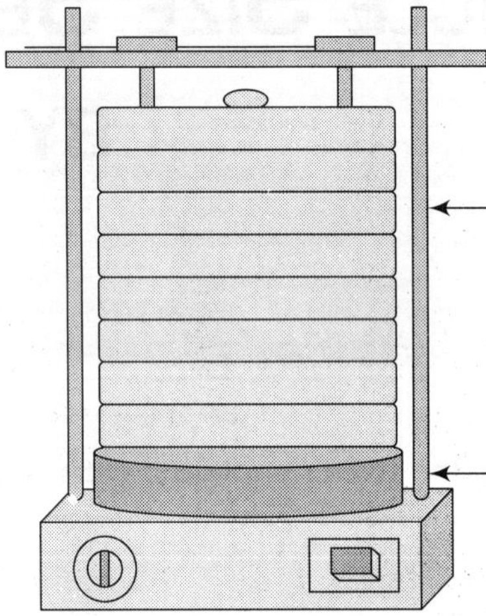

Fig. 17.1: Sieve assemble with Oscillator.

PROCEDURE

1. 50 gm of the sample ($CaCO_3$) is weighed.
2. The sieves are arranged in order of decreasing size of the openings in the mesh with the coarsest at the top.
3. Place the assembly of the sieves on the mechanical sieve shaker.
4. Put the weighed sample on the top sieve.
5. Close the sieves assemble with the lid. and tighten the screws so that the assemble is firmly present over the sieve shaker.
6. Set the optimum (medium) speed.
7. Switch the power on.
8. Switch on the mechanical sieve shaker.
9. Allow it to be in motion for 20 min.
10. After 20 minute the sieves are taken out and find out the weight of the particle retained over each sieves.
11. The valves are tabulated.
12. Find out the percentage of the powder retained on each sieve.
13. Find out the arithmetic mean.
14. Calculate % weight of powder over size by adding up the values of % of weight of powder retained.

15. % under size is calculated by subtracting the values of % weight of powder retained from 100.
16. Plot the corresponding graphs.

OBSERVATIONS AND CALCULATIONS

Table 17.1: Determination of Particle Size and Distribution

Sl No	** Sieve opening in mm*	*Log of screen opening*	*Weight of powder retained in gm*	*% wt of powder retained*	**Arithmetic mean size opening(mm)*	*Weight size*	*Arithmetic mean*	*Cumulative % weight over size*	*Cumulative % weight under size*
1				–				0	100
2				x				$0 + x$	$100 - x$
3				y				$0 + x + y$	$(100 - x)$
4									$- y$
5									
6									
7									
8									
9									
10									

* Refer to Appendix VIII

% weight of powder retained = Weight of powder retained / 50 X 100
= ....

Weight size = Arithmetic mean sieve opening X % Weight of powder retained
= ....

Arithmetic mean = Weight size/100
= ....

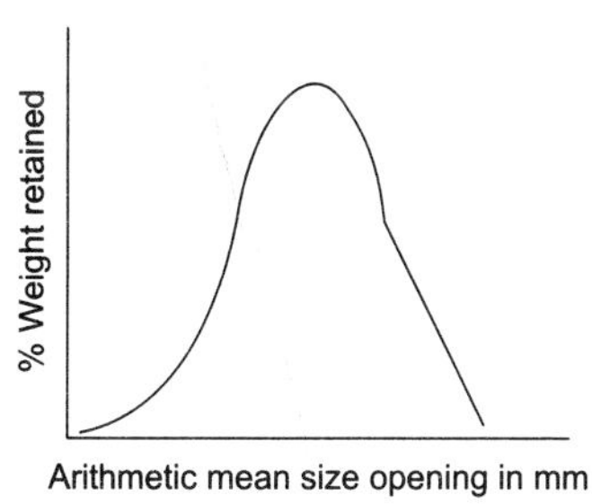

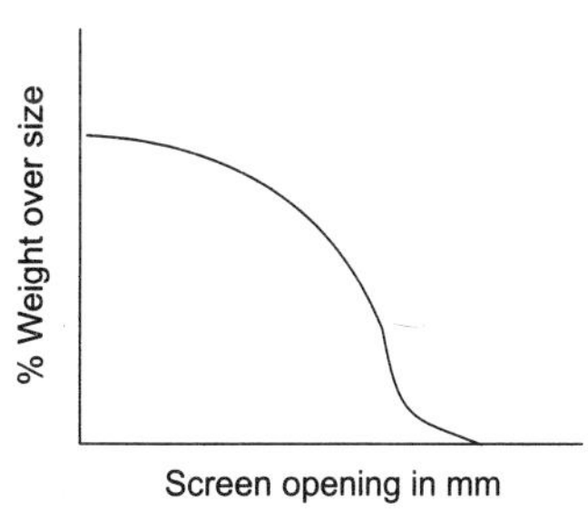

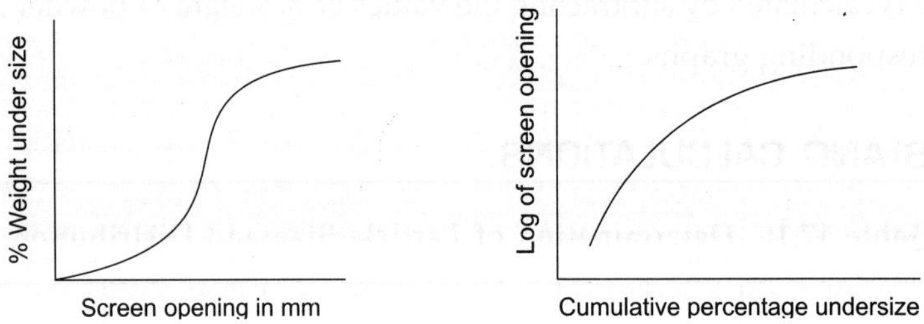

Fig. 17.2: Size distribution of particles present in the sample.

REPORT

Size separation of the given sample is performed.

The size distribution is shown in the form of graphs.

The average particle size of the given sample is .... mm.

EXPERIMENT 18

PARTICLE SIZE SEPARATION — BEAKER DECANTATION

AIM

To separate particles of known size from the prepared suspension.

REQUIREMENTS

Beakers	(4)
Motor and pestle	(1)
Glass rod	(4)
Measuring cylinder	(2)
China dish	(1)
Stopwatch	(1)
Weighing Balance	(1)
$CaCO_3$	
Water	

PRINCIPLE

During sedimentation the particles will accelerate under gravity. As it accelerates the fluid will offer some resistance to the sedimenting particles. At a particular time this resisting force becomes exactly equal to the force of gravity. And from this time onwards the particle will starts to settle with a constant velocity. This is called as the terminal settling velocity. The terminal velocity of particles can be determined by stocks law,

$$V = \frac{d^2(\rho_s - \rho_o)g}{18\eta} \quad \text{...(Eq. 18.1)}$$

where,

d = Diameter of the particle (m)

ρ_s = Density of the particle (g/cc)

ρ_o = Density of liquid (g/cc)

g = Acceleration due to gravity (m/s^2)

From this equation it is possible to determine the terminal velocity of particles of known diameter. Say suppose particle of size 40 µm, 20 µm etc.

If a suspension taken in a container of known volume, let the height of the suspension in the container is H_o. It is possible to determine the velocity of settling,

$$\boxed{V = H_o/t} \quad \text{... (Eq. 18.2)}$$

After rearranging the equation one can find out the settling time for particles of interest. By applying this concept it is easy to separate particles of different size by pre calculating the time for them to settle completely. This experiment allows an easy mode of separation of particle of interest by performing simple decantation procedure.

PROCEDURE

1. Prepare slurry of calcium carbonate in water (5 g/100 ml).
2. Add this slurry into a beaker (I) and make u to the mark depending upon the capacity of the beaker.
3. Note down the initial height of the slurry.
4. With the help of the equation, determine the velocity of settling for different particles of diameter 40 µm, 20 µm.
5. The corresponding time to settle the particles is also determined. (For *e.g.*, 40 µm and above particles settle completely at X min, and 40 – 20 µm at Y min).
6. Immediately before starting the experiment stir the suspension in the beaker with a glass rod.
7. Start the stop watch immediately after stirring.
8. Allow the slurry to settle down.
9. After completion of X time stop the watch and carefully transfer the supernatant liquid to another beaker (II) of same capacity without disturbing the sediment.
10. Transfer the sediment into a china dish (previously weighed).
11. Dry the particles of the sediment obtained from beaker (I) in an oven.
12. Find out the weight of the particles (particle size will be 40 µm and above)
13. Make up the volume of the second beaker with water to the original initial height.
14. Repeat the step 6, 7, & 8.
15. After completion of Y time carefully transfer the supernatant liquid to another beaker (III) of same capacity without disturbing the sediment.
16. Repeat the steps 10 & 11.
17. This will contain the particles of size 20 – 40 µm.
18. Transfer the supernatant into a china dish. Evaporate the water and find out the weight of the particle present in the supernatant.
19. This will contain particles of size below 20 µm.
20. Find out the weight of dried residue in the china dish.

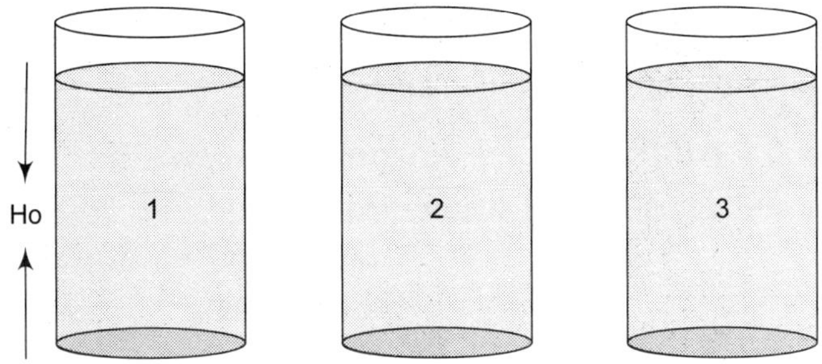

Fig. 18.1: Sedimentation of particles of different size present in the same sample of $CaCO_3$.

Beaker 1 – Sediment containing particle size > 40 μm

Beaker 2 – Sediment containing particle size 40 – 20 μm

Beaker 3 – Sediment containing particle size < 20 μm

OBSERVATIONS AND CALCULATIONS

Initial weight of the $CaCO_3$ = 5 gm

Initial height of the slurry in the beaker = 20 cm

Specific gravity of $CaCO_3$ = 2.5 g/ cc

Specific gravity of water = 1 g/cc

Viscosity of water = 0.1 poise

Acceleration due to gravity = 980 cm/s^2

Table 18.1: Determination of Settling time of Particles of Different Size

Sl No	*Particle size (μm)*	*Initial height of the suspension (cm)*	*Settling velocity (cm/sec)*	*Settling time (min)*
1	40 and above			
2	40-20			
3	Below 20			

Table 18.2: Determination of % Weight of Particles Settled

Sl No	*Size (μm)*	*Settling time (min)*	*Weight of empty china dish (gm)*	*Weight of china dish + sediment (after drying) (gm)*	*Weight of the residue (gm)*	*Percentage weight = wt of the* $\frac{\text{Residue}}{5} \times 100$
1	Above 40 μm					
2	40 μm-20 μm					
3	Less than 20 μm					

$$\text{Velocity of settling for particle size } 40\ \mu\text{m} = \frac{\left(40\times10^{-4}\right)^2(2.5-1)980}{18\times0.1}$$

$$= 0.013067 \text{ cm/ sec}$$

$$\text{Time for settling particles of } 40\ \mu\text{m} = \text{Ho/ V}$$

$$= \ldots../\ 0.013067 = \ldots\ldots\ldots \text{ sec}$$

$$= \ldots\ldots\ldots\ldots \text{min}$$

$$\text{Velocity of settling for particle size } 20\ \mu\text{m} = \frac{\left(20\times10^{-4}\right)^2(2.5-1)980}{18\times0.1}$$

$$= 0.00326 \text{ cm/sec}$$

$$\text{Time for settling particles of } 20\ \mu\text{m} = \ldots../\ \text{V}$$

$$= \ldots../\ 0.00326$$

$$= \ldots\ldots \text{ sec}$$

$$= \ldots\ldots.\text{min}$$

REPORT

Particles of different size are separated by simple decantation method.

EXPERIMENT 19

PARTICLE SIZE SEPARATION BY SEDIMENTATION METHOD

AIM

To separate the particle by sedimentation method.

To determine the stock's diameter, and other statistical diameters of the separated particles.

REQUIREMENTS

Andreasen apparatus,	(1)
Beaker 1000 ml	(1)
Stopwatch	(1)
Conical Flask	(4)
China dish	(4)
Hot air oven	(1)
Weighing balance	(1)
Weighing box	(1)
Long Scale	(1)
Water bath	(4)
Tween 80	
Calcium carbonate	

PRINCIPLE

Particle size separation can be achieved by a number of methods like screening, sedimentation, air separation, and elutriation methods. If the particles are too small to screen effectively or the amount to be separated is more, then sedimentation method is the advised simple method. In a simple sedimentation, separation is happened under gravity depending upon the size as well as the density of the particles. So sedimentation method can be used for determining the size of the particle settled.

Several methods are available based on sedimentation for the size separation and particle size determination. They are hydrometer method, pipette method and balance method. Andreasen apparatus works under the principle of pipette method. The particle size in the sub sieve range may be obtained by sedimentation as expressed in Stoke's law,

$$V = h/t = \frac{d_{st}^2(\rho_s - \rho_0)g}{18\eta_0} \quad \text{... (Eq. 19.1)}$$

Or

$$d_{st} = \sqrt{18\eta_0 h/(\rho_s - \rho_0)gt} \quad \text{... (Eq. 19.2)}$$

where,

V = Rate of settling (m/s)

h = Distance of fall within time (t)

d_{st} = Mean diameter of particle based on the velocity of settling

ρ_s/ρ_0 = Density of particle and water (kg/m^3)

η_0 = Viscosity of medium (Poise)

The particle diameter corresponding to various time intervals is calculated from Stoke's law. The residue or dried sample obtained at a particular time is the weight fraction having particle of size less than that size obtained from Stoke's law. Hence each sample is therefore called as weight undersize. Add up the successive weight undersize (expressed in %). This is then plotted on log probability scale (log normal) against particle diameter (log scale) and the geometrical mean diameter d'g can be determined from the reference point *i.e.,* 50% on the probability scale and geometrical standard deviation σg can be determined from the slope. The other statistical parameters can be determined from the equations,

$\log d_{ln}$ (Length number mean diameter) = $\log d'g - 5.757 \log^2 \sigma g$

$\log d_{sn}$ (Surface number mean diameter) = $\log d'g - 4.606 \log^2 \sigma g$

$\log d_{vn}$ (Volume number mean diameter) = $\log d'g - 3.454 \log^2 \sigma g$

$\log d_{vs.}$ (Volume surface mean diameter) = $\log d'g - 1.151 \log^2 \sigma g$

$\log d_{wm}$ (Weight momentum mean diameter) = $\log d'g + 1.151 \log^2 \sigma g$

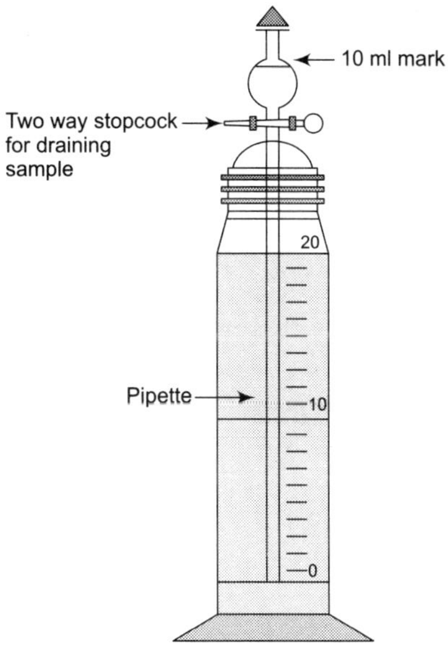

Fig. 19.1: Andreasen apparatus.

PROCEDURE

1. Prepare a 600 ml of 0.2% Tween 80 solution with distilled water.
2. Dissolve 6 grams of calcium carbonate and form a smooth paste and mix with the total quantity of Tween 80 solution.
3. This suspension is transferred to the andreasen apparatus up to the 550 ml mark.
4. Close the stopper and assure that the lower tip of the pipette is 20 cm below the surface of the liquid (suspension).
5. The suspension is agitated with a plunger type stirrer. And start the stopwatch.
6. Immediately after agitation 10 ml of the sample is taken by sucking through the pipette and transferred into a conical flask. Do not cause any disturbance to the apparatus.
7. Transfer it into a china dish. (Which is previously weighed W_1).
8. The sample is evaporated with the help of water bath. And dried in a oven.
9. Find out the weight of the residue + china dish (W_2).
10. The weight of the residue can be determined by $W_2 - W_1$.
11. After 2 min the second sample is withdrawn without disturbing the sediment if formed.
12. The step 7, 8, 9, 10 is repeated.
13. Samples are withdrawn at an interval of 2 min up to 30 min.
14. The weight of the particle residue at each 2 min is determined.
15. After taking each 10 ml from the apparatus the height of the liquid column is noted. (Height from the tip of the pipette to the surface of the liquid.) For the 0 min the height will be 20 cm and for the succeeding it will decrease.
16. Find out the stoke's diameter of particle settled in each 2 min with the help of equation.
17. Find out the % weight under size, $\frac{\text{Weight obtained}}{\text{Total weight of powder present in the sample}} \times 100$ and cumulative weight under size (by adding up the values obtained in the % weight under size)

18. Draw the weight distribution curve by taking stokes's diameter in X axis and corresponding wt under size on Y axis.
19. Draw another graph by taking log stoke's diameter on X axis and cumulative % undersize on Y axis (use log normal graph/Semi logarithmic graph).
20. Determine the geometrical mean diameter d'g from the graph by finding out the logarithm of particle size equivalent to 50% on the probability scale, *i.e.*, the 50% size.
21. Calculate the standard deviation by finding out the slope of the graph. Using the equation, 50% size/ 16% undersize or 84% over size.
22. With the help of the equations provided find out the other statistical parameters.

OBSERVATIONS AND CALCULATIONS

$$\text{Stoke's diameter } d_{st} = \sqrt{18\eta_0 h / (\rho_s - \rho_0) gt}$$
$$= \sqrt{18 \times 0.01 \times \ldots\ldots / (0.0025 - 0.00098) \times 9.8 \times \ldots\ldots}$$

V = Rate of settling m/s

h = Distance of fall within time t in sec

d_{st} = Mean diameter of particle based on the velocity of settling

ρ_s = Density of $CaCO_3$ (0 .0025 kg/m^3)

ρ_0 = Density of particle and water (0.00098 in kg/ m^3)

η_0 = Viscosity of medium (0 .01 poise)

log d'g = ………………..

σ g = ………………..

Table 19.1: Determination of Weight of the Particles After Separation

Sl No	*Time in min*	*Weight of empty china dish* W_1 *(gm)*	*Weight of china dish +residue after drying* W_2 *(gm)*	*Weight of the residue* W_3 *(gm)*
1	0			
2	2			
3	4			
4	6			
5	8			
6	10			
7	12			
8	14			
9	16			
10	18			
11	20			
12	22			
13	24			
14	26			
15	28			
16	30			

Table 19.2: Determination of Particle Size

Sl No	*Time (min)*	*Height of the liquid column (m)*	*Weight of the residue (gm)*	*% weight under size*	*Cumulative weight under size*	*Stoke's diameter d_{st}*
1	0					
2	2×60					
3	4×60					
4	6×60					
5	8×60					
6	10×60					
7	12×60					
8	14×60					
9	16×60					
10	18×60					
11	20×60					
12	22×60					
13	24×60					
14	26×60					
15	28×60					
16	30×60					

$\log d_{ln}$ (length number mean diameter) $= \log d'g - 5.757 \log^2 \sigma g$

$= \ldots\ldots\ldots\ldots\ldots\ldots$

$\log d_{sn}$ (surface number mean diameter) $= \log d'g - 4.606 \log^2 \sigma g$

$= \ldots\ldots\ldots\ldots\ldots\ldots\ldots$

$\log d_{vn}$ (volume number mean diameter) $= \log d'g - 3.454 \log^2 \sigma g$

$= \ldots\ldots\ldots\ldots\ldots\ldots\ldots$

$\log d_{vs.}$ (Volume surface mean diameter) $= \log d'g - 1.151 \log^2 \sigma g$

$= \ldots\ldots\ldots\ldots\ldots\ldots\ldots$

$\log d_{wn}$ (Weight momentum mean diameter) $= \log d'g + 1.151 \log^2 \sigma g$

$= \ldots\ldots\ldots\ldots\ldots\ldots\ldots$

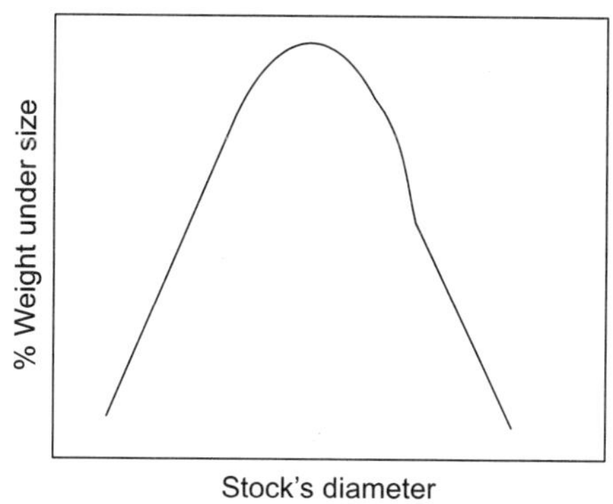

Fig. 19.2: Particle size distribution.

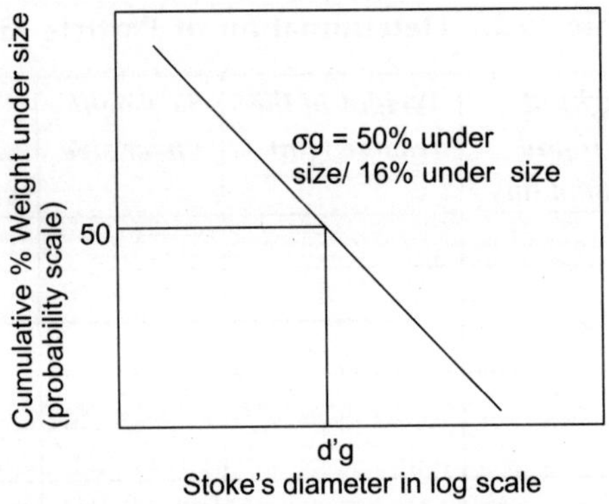

Fig. 19.3: Log probability plot.

REPORT

The Stoke's diameter of particle depending upon their sedimentation velocity is determined

The geometrical mean size of the sample is d'g =

The geometrical standard deviation σg =

The length number mean diameter d_{ln} =

The surface number mean diameter d_{sn} =

The volume number mean diameter d_{vn} =

The volume surface mean diameter d_{vs} =

The weight moment mean diameter d_{vm} =

EXPERIMENT 20

MIXING – DETERMINATION OF UNIFORMITY INDEX

AIM

To perform the mixing of lactose with salicylic acid.

To determine the uniformity of mixing by uniformity index.

To determine the time taken for perfect mixing.

REQUIREMENTS

Cone blender	(1)
Colorimeter	(1)
Test tubes	(10)
Volumetric flasks (100ml)	(6)
Weighing balance	(1)
Weight box	(1)
Butter paper	
Filter paper	
Lactose, Salicylic acid, Ferric chloride (4%)	

PRINCIPLE

Mixing is defined as an operation in which two or more components in a separate condition are treated so that each particle lies as nearly as possible in contact with particles of each of the other ingredients. The uniformity index is a marker of effectiveness of mixing operation.

During mixing the two major problems arises are:

1. How long the mixing should be proceeded to obtain the uniformity
2. Whether the uniformity is achieved or not.

These two problems can be rectified with the determination of uniformity index. The criterion for uniformity of mixing can be determined by the equation,

$$M_c = \left[\frac{\Sigma(C_x - C_{av})^2}{NC_{av}^2}\right]^{0.5} \quad \text{... (Eq. 20.1)}$$

where,

M_c = The criterion for uniformity of mixing
C_x = The amount of any one constituent present in the sample collected
C_{av} = The average amount of the any one constituent present in the selected samples at particular time interval
n = Number of samples

The criterion of unmixed batch,

$$M_{co} = \left[\frac{1 - C_{av}}{C_{av}}\right]^{0.5} \quad \text{... (Eq. 20.2)}$$

The uniformity index

$$I = M_c/M_{co} \quad \text{... (Eq. 20.3)}$$

The uniformity index varies from one to zero. As the value approaches the zero it is concluded that perfect mixing is achieved.

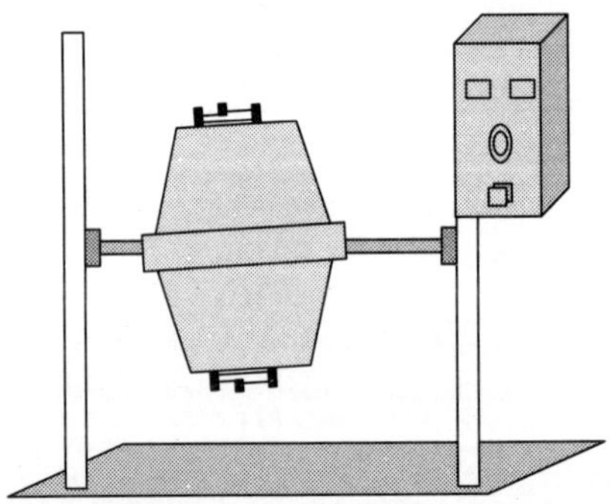

Fig. 20.1: Double cone blender.

PROCEDURE

1. Clean the blender (Cone blender).
2. Weigh 100 gm of lactose and 2 gm of salicylic acid.
3. Add carefully in to the blender.
4. Close the lid of the blender tightly.
5. Make sure that no leakage of material is happening during mixing.
6. Switch on the power supply.
7. Allow to rotate the cone blender on its medium speed.
8. Switch OFF the blender after 15 min.

9. Take out 3 samples each of approximately 500 mg from different parts of the blender. (sample from centre, left, right)
10. Carefully transfer each sample into different volumetric flasks (100 ml).
11. Add water and shaken until the salicylic acid is completely dissolved.
12. Make up the volume into 100 ml.
13. Label the corresponding samples as 15 centre, 15 left, and 15 right.
14. Close the blender and allow to rotate for another 15 min.
15. Repeat the steps 8-12
16. Label the samples as 30 centre, 30 left, and 30 right.
17. Allow to rotate one more time For 15 min.
18. And the same procedure is repeated up to 60 min.

    Analysis of the sample

    Pipette out 10 ml of sample from the volumetric flask to a 25 ml test tube and mark accordingly. Add 5 ml of ferric chloride solution 4% w/v. A pinkish violet color developed. Make up the volume to 20 ml with distilled water. The density of the color developed can be analyzed with the help of colorimeter. (Refer to Appendix IX)
19. Determine the concentration of salicylic acid present in each sample.(See the graph Fig. 20.2)
20. The values are tabulated and uniformity index for different time intervals are determined.
21. The experiment is repeated until the uniformity index is approximately reaches zero.

OBSERVATIONS AND CALCULATIONS

Table 20.1: Colorimetric Determination of Salicylic Acid

Sl No	*Time (min)*	*Samples*	*Absorbance*	*con /ml*	*con/100ml (C_x)*
1	15	Centre			
		Left			
		Right			
2	30	Centre			
		Left			
		Right			
3	45	centre			
		Left			
		Right			
4	60	centre			
		Left			
		Right			

Table 20.2: Determination of Uniformity Index

Sl No	*Time*	*Sample*	*con/100ml (C_x)*	*Average con (C_{av})*	$C_x - C_{av}$	$(C_x - C_{av})^2$
1	15	Centre				
		Left				
		Right				
2	30	Centre				
		Left				
		Right				
3	45	Centre				
		Ceft				
		Right				
4	60	Centre				
		Left				
		Right				

The criterion of the unmixed batch $(\mathbf{M}_{co}) = \left[\frac{1 - C_{av}}{C_{av}}\right]^{0.5} = 1$

The criterion of mixing $(\mathbf{M}_C) = \left[\frac{\Sigma(C_x - C_{av})^2}{NC_{av}^2}\right]^{0.5}$

At different time interval, **N** = The number of samples taken after each interval

= ....

The uniformity index

$$I = M_c / M_{co}$$

= ....

Table 20.3: Determination of Change of Uniformity Index with Time

Sl No	*Time in min*	*Uniformity index(I)*
1	15	
2	30	
3	45	
4	60	

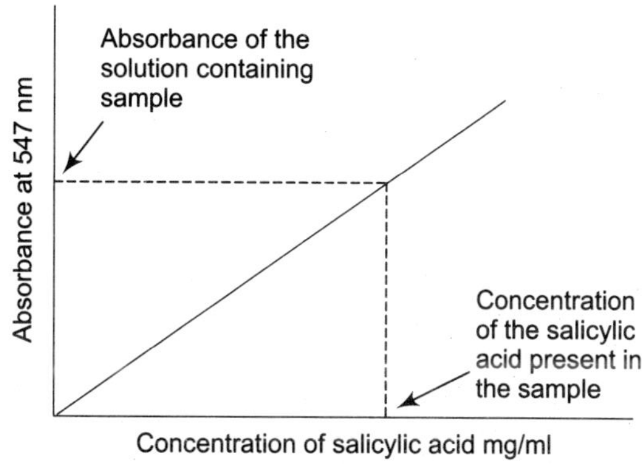

Fig. 20.2: Determination of unknown concentration of salicylic acid.

REPORT

The uniformity index at 15 min I (15) = ...

I (30) = ...

I (45) = ...

I (60) = ...

The complete mixing is achieved at .... min

EXPERIMENT 21

LIQUID-LIQUID MIXING—DETERMINATION OF UNIFORMITY OF MIXING

AIM

To determine the uniformity of mixing (liquid / liquid) by specific gravity method.

REQUIREMENTS

Mechanical stirrer	(1)
Beaker (1000 ml)	(1)
Pipette (20 ml)	(1)
Measuring cylinder (100 ml)	(1)
Specific gravity bottle	(3)
Water	
Glycerin	

PRINCIPLE

Mixing is defined as an operation in which two or more components in a separate condition are treated so that each particle lies as nearly as possible in contact with particles of each of the other ingredients. The uniformity index is a marker of effectiveness of mixing operation.

During mixing the two major problems arises are,

1. How long the mixing should be proceeded to obtain the uniformity.
2. Whether the uniformity is achieved or not.

These two problems can be rectified with the determination of uniformity index. Unlike the solid mixing the determination of uniformity of mixing in liquid-liquid mixing is difficult. The simplest method is by the determination of specific gravity of the samples taken from different places of container while mixing. By knowing the original density of both the liquids one can easily identify the change in density with the help of specific gravity bottle. If the liquids are not properly mixed it will show large variation among the samples. Complete mixing can be ensured when all the samples taken from different parts of the container shows same specific gravity.

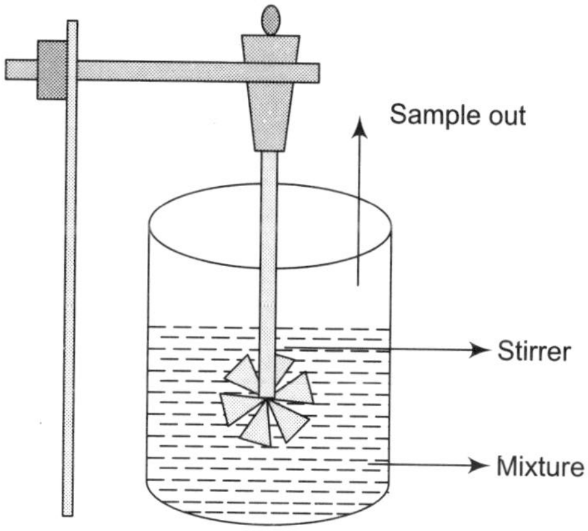

Fig. 21.1: Arrangement for determination of uniformity of liquid-liquid mixing.

PROCEDURE

1. Clean the mechanical stirrer.
2. Take 750 ml of water in a 1000 ml beaker.
3. Add 50 ml of glycerin to it.
4. Place the stirrer inside the beaker at the center portion (½ amount of the liquid should present above the leaves of the stirrer).
5. Set a minimum speed of rotation for the stirrer.
6. Allow the stirrer to rotate at the pre set speed.
7. Switch off the stirrer after 5 min.
8. Pipette out the samples 20 ml each from 3 different places of the beaker (Preferably from the centre, top, and bottom).
9. Find out the specific gravity of the samples. (values are preferred up to 4 decimals) (Refer to appendix IV).
10. Replace the samples to the beaker and allow the stirrer to rotate another five min and repeat the steps 7-9.
11. Experiment is repeated until all the samples taken from different places show same specific gravity.

OBSERVATIONS AND CALCULATIONS

Determination of Density

Weight of the empty specific gravity bottle $(\mathbf{W_1})$ = ... gm

Weight of the bottle + water $(\mathbf{W_2})$ = ... gm

Weight of the bottle + sample $(\mathbf{W_3})$ = ... gm

$$\rho = W_3 - W_1 / W_2 - W_1$$

Table 21.1: Determination of Specific Gravity of Mixture

Sl No	*Time in min*	*Samples*	*Specific gravity (gm/ml)*
1	5	Centre	
		Surface	
		Bottom	
2	10	Centre	
		Surface	
		Bottom	
3	15	Centre	
		Surface	
		Bottom	
4	20	Centre	
		Surface	
		Bottom	
5	25	Centre	
		Surface	
		Bottom	
6	30	Centre	
		Surface	
		Bottom	
7	35	Centre	
		Surface	
		Bottom	
8	40	Centre	
		Surface	
		Bottom	

REPORT

The complete Mixing is achieved with in .... min

The specific gravity of the completely mixed liquid is .... gm /ml

EXPERIMENT 22

DETERMINATION OF CRITICAL POINT OF SETTLING—SEDIMENTATION

AIM

To determine the critical point of settling for the given suspension of drug.

REQUIREMENTS

Measuring cylinder 100 ml	(1)
Glass rod	(1)
Motor and pestle	(1)
Beaker 250 ml	(1)
Scale	(1)
Marker pen	(1)
Stopwatch	(1)
$CaCO_3$	

PRINCIPLE

If a suspension is taken in a long glass cylinder, and kept undisturbed its starts settling purely due to gravity. During settling an observer can see the following zones (See Fig. 22.1). During the start of the experiment there is uniform concentration of drug throughout the cylinder. The coarser particle will settle immediately depending upon the size, first forming a layer of settled coarse particles (E) at the bottom of the cylinder. Next forms a layer of settling particles called as compression zone (D) with a transition zone of partially thickened layer above it (C). The next zone having original concentration (B) and the top most layer (A) is the clear water. As time proceeds changes occurred for interface between zone A & B, and C & D. The time at which the complete disappearance of B & C zone is occurred is considered as the critical point of settling.

A plot by taking time in minutes at X axis and height in inches or cm on Y axis can be constructed. The upper curve showing the position of interface between zone A & B and lower curve showing the poison of the interface between zone C & D with time. The critical point is directly found out from the plot where the two curves unite.

PROCEDURE

1. Prepare slurry of calcium carbonate in water (5 g/100 ml).
2. Add this slurry to a measuring cylinder of 100 ml capacity.
3. Immediately before starting the experiment stir it with a glass rod.
4. Start the stopwatch immediately after stirring.
5. Allow the slurry to settle down.
6. At an interval of 10 min note down the height of the interface zone of A & B and C & D. (height is measured from the bottom of the cylinder to the interface)
7. Experiment is proceeded till B and C layer disappears. (at this point only the D and A layer will be present)

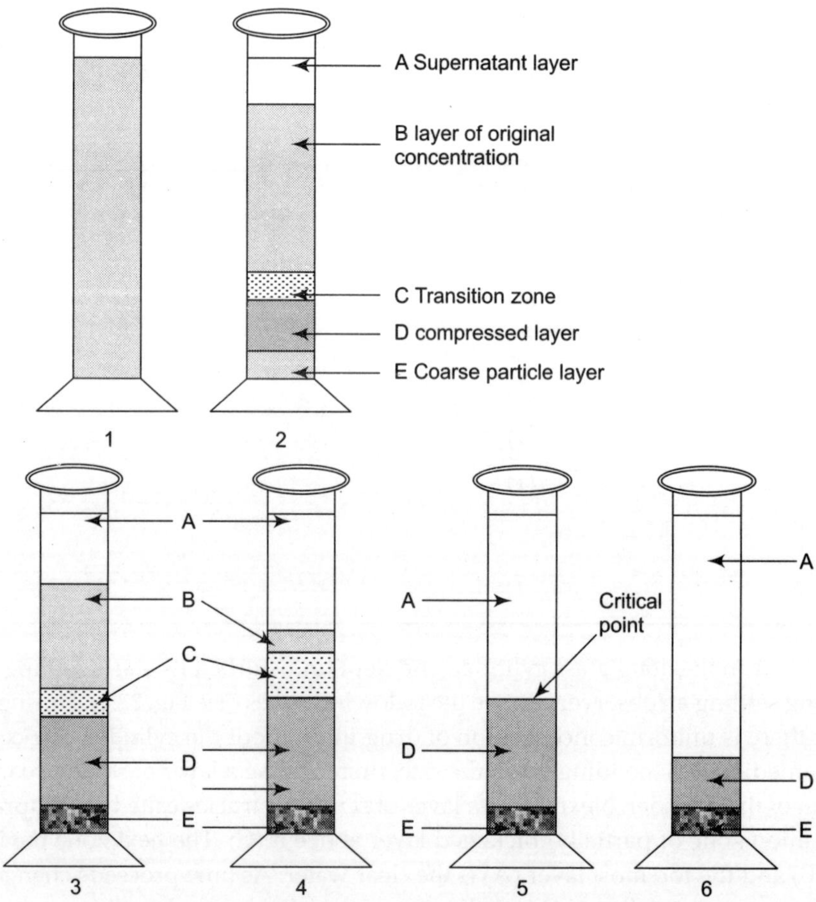

Fig. 22.1: Changes occurring during sedimentation.

8. Plot the upper curve and lower curve by taking time on X axis and height in cm on Y axis.
9. Mark the point where the two curves coincide.

OBSERVATIONS AND CALCULATIONS

Table 22.1: Determination of Sedimentation with Time

Sl No	*Settling time(min)*	*Height of the interface between A & B from the bottom (cm)*	*Height of the inter face Between C & D from the bottom (cm)*
1	10		
2	20		
3	30		
4	40		
5	50		
6	60		
7	70		
8	80		
9	90		
10	100		
11	110		
12	120		

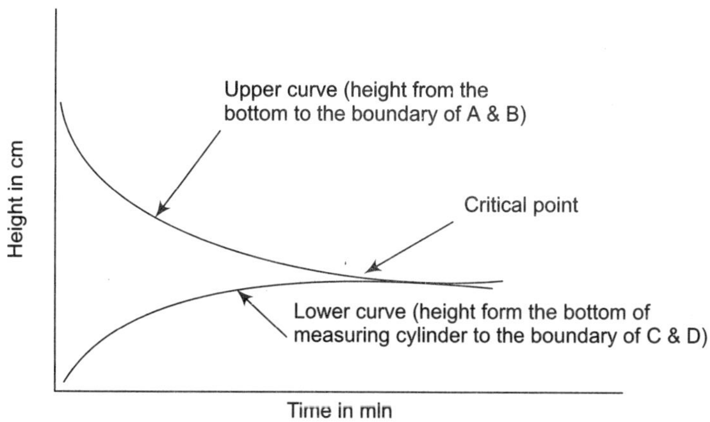

Fig. 22.2: Determination of critical point of sedimentation.

REPORT

The critical point of settling is achieved at ——————————————— min.

EXPERIMENT 23

SEDIMENTATION—EFFECT OF SEDIMENTATION RATE ON CONCENTRATION

AIM

To demonstrate the effect of sedimentation rate on the concentration of the capacity limiting zone.

REQUIREMENTS

Measuring cylinder 100 ml	(1)
Glass rod	(1)
Motor and pestle	(1)
Beaker 250 ml	(1)
Scale	(1)
Marker	(1)
Stopwatch	(1)
$CaCO_3$	

PRINCIPLE

During sedimentation different zones are developed as discussed in the previous experiment. In batch settling the position as well as concentration of the zones varies with time. The concentration at the zone B (capacity limiting zone) with time can be determined by the following equation,

$$\boxed{C_L = \frac{C_o Z_o}{Z_i}} \qquad \text{... (Eq. 23.1)}$$

where,

Z_i = The height of the uniform slurry of concentration C_L (Can be determined from the graph of height of the interface between the clear liquid and zone B on Y axis and time on the X axis.)

C_0 = Initial concentration of the slurry

Z_0 = Initial height of the slurry

The concentration of zone B at different time intervals and the corresponding velocity of settling can be determined. This experiment is providing a basic knowledge for the determination of area needed for proper sedimentation in a continuous thickener.

PROCEDURE

1. Prepare slurry of calcium carbonate in water. (5g/100 ml)
2. Add this slurry to a measuring cylinder of 100 ml capacity. (The inner diameter of the cylinder should be noted)
3. Immediately before starting the experiment stir it with a glass rod.
4. Start the stopwatch immediately after stirring
5. Allow the slurry to settle down.
6. At an interval of 10 min note down the height of the interface zone of A & B and C & D. (Height is measured from the bottom of the cylinder to the interface)
7. Experiment is preceded till B and C layer disappears. (at this point only the D and A layer will be present)
8. Plot the upper curve by taking height of the interface (A & B) in inch or cm on Y axis and time on X axis.
9. From Arbitrarily chosen values of time tangents are drawn to the Y axis. The tangent intercepts the vertical axis is considered as the Zi for that time of sedimentation.
10. Find out the slop of the tangent. This will give the settling velocity at the capacity limiting zone. (Rate of settling in inch or cm/min)
11. Concentration of the zone can be determined with the help of equation 23.1.
12. Tabulate the values obtained.
13. Draw another graph by taking concentration of the capacity limiting zone (gm/ml) on X axis and settling velocity on Y axis.

OBSERVATIONS AND CALCULATIONS

Table 23.1: Determination of C_L

Sl No	*Time (min)*	*Height of the interface between A&B from the bottom (cm)*	*Z_i (cm)*	*Settling velocity V_L (cm /min)*	*Concentration of the zone B* $C_L = \frac{C_0 Z_0}{Z_i}$ *(g/ml)*
1	10				
2	20				
3	30				
4	40				
5	50				
6	60				

7	70				
8	80				
9	90				
10	100				
11	110				
12	120				

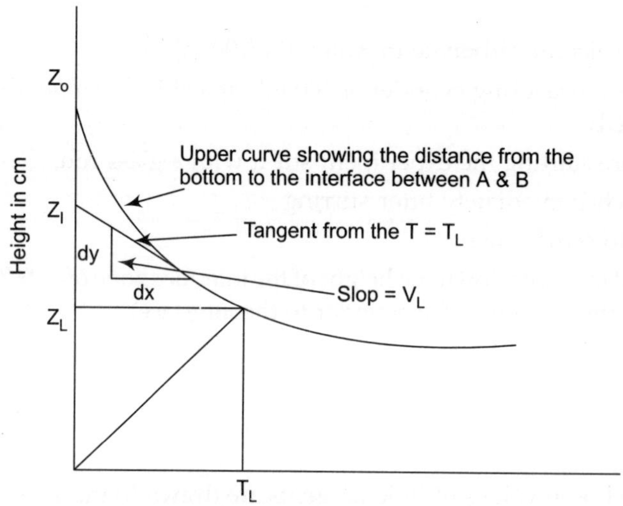

Fig. 23.1: Determination of velocity of settling.

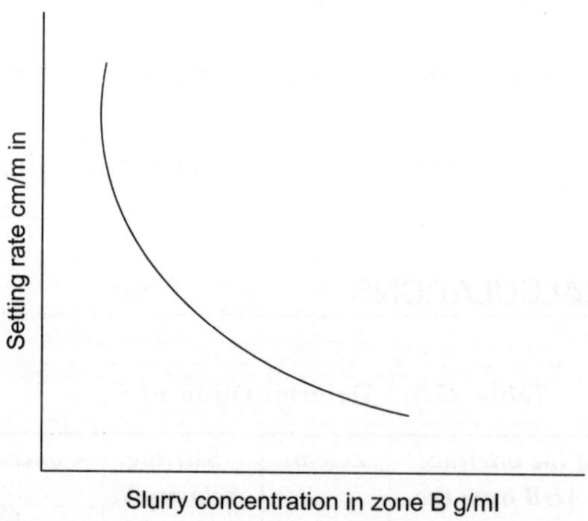

Fig. 23.2: Effect of settling velocity on concentration.

REPORT

Settling velocities at different time intervals are determined and studied the effect of settling velocity on concentration of drug at zone B (capacity limiting zone).

EXPERIMENT 24

DETERMINATION OF TEMPERATURE HUMIDITY INDEX—THI

AIM

To determine the temperature humidity index of the atmosphere.

REQUIREMENTS

Wet bulb and dry bulb thermometers (Mansion type) (1)

Water

PRINCIPLE

Temperature humidity index is a number that express the degree of discomfort of the average individual due to the atmospheric temperature and relative humidity.

If the atmospheric temperature increases, the flow of heat from the body to the surroundings decreases. Increased humidity is also a reason for the lowering the air flow from body to atmosphere (because of the slow evaporation of sweat from the body). So the temperature and humidity determines the human comfort. If the humidity is high the feel of "hot" is more compared to the less humid area of same dry bulb temperature. THI can be determined by the following equation,

THI = (0.4 X Wet bulb temperature + Dry bulb temperature) + 15

THI	Atmosphere
<70	Comfortable
70 -80	Causes discomfort to some individuals
>80	Causes discomfort to all human beings

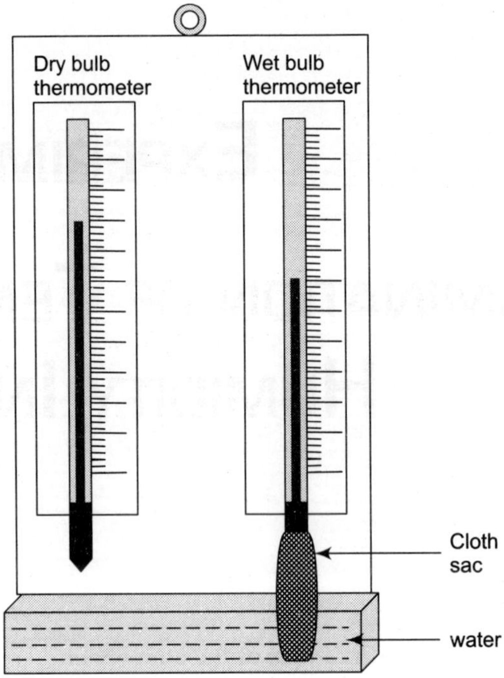

Fig. 24.1: Arrangement of wet bulb and dry bulb thermometer.

PROCEDURE

1. Arrange the wet bulb and dry bulb thermometers. One thermometer is exposed to the air (dry bulb thermometer) and the bulb of the second is covered with a cloth sac (wet bulb thermometer).
2. Place water in the plastic container, and the cloth sack is dipped into it. So that by capillary action the water will reach the bulb of the wet bulb thermometer and wet the bulb.
3. Carefully watch the readings in both the thermometer.
4. The reading in the wet bulb thermometer decreases as it is in direct contact with moisture content.
5. After some time the reading in the wet bulb thermometer reaches an equilibrium constant temperature. Note down this reading.
6. Simultaneously note down the dry bulb temperature.
7. Take the 2 more readings after a time interval of 5 min
8. Substitute in the equation and find out the temperature humidity index.

OBSERVATIONS AND CALCULATIONS

Table 24.1: Determination of THI

Sl No	*Dry bulb Temperature °F*	*Average dry bulb temp °F*	*Wet bulb temperature °F*	*Average Wet bulb temp °F*	*Wet bulb temp + Dry bulb temp*	*THI*
1						
2						
3						

THI = (0. 4 × Wet bulb temperature + Dry bulb temperature) + 15

= ....

REPORT

The Temperature humidity index of the atmosphere is found to be =

This atmosphere is said to be .... to human beings

EXPERIMENT 25

DETERMINATION OF HUMIDITY BY PSYCHROMETRIC METHOD

AIM

To determine the humidity of the atmosphere by psychrometric method.

To determine the dew point, relative humidity, humid heat, and humid volume by psychrometric method.

To determine the level of human comfort of the atmosphere.

REQUIREMENTS

Wet bulb and dry bulb thermometers (Manson's type)	(1)
Psychrometric chart (Humidity Chart)	(1)
Water	

PRINCIPLE

Water exists in the air at all temperatures. It is always in the form of vapour. The quantity of the water vapour contained in the air occupies a very small part of air volume. The heat content of the air is the heat contained in the water vapour. Water vapour would condense when it comes in contact with any material at lower temperature.

The quantity of water vapour present in the atmosphere is not constant as evaporation is continuously taking place from the surface of water in lakes, river and ocean and even from the human body. The quantity of the heat offered by the atmosphere is depends upon these evaporations. Instead of the continuous evaporation taking place from the earth the air never become saturated. There is a maximum limit to the amount of water vapour that air can hold (saturated air).

Humidity can be defined as the pounds of water vapour carried by 1 lb of dry air at a particular temperature. The humidity can be recorded in terms of lb of water vapour/lb of dry air. Relative humidity is a term used to describe the amount of water vapour that exists in a gaseous mixture of air and water vapour, compared to the upper limit of what it could be at the same temperature and bulk pressure. Determination of relative humidity is important for climate control (HVAC system) of building, vehicle etc. it also determine the human comfort to the atmospheric conditions. The following table gives an idea about how the dew point and relative humidity affects the human comfort.

Dew Point °C	*Dew Point °F*	*Human Perception*	*Rel. Humidity at 32 °C (90 °F)*
> Higher than 26 °C	>Higher than 80 °F	Severely high. Even deadly for asthma related illnesses	65% and higher
24 – 26 °C	75 – 80 °F	Extremely uncomfortable, fairly oppressive	62%
21 – 24 °C	70 – 74 °F	Very humid, quite uncomfortable	52% – 60%
18 – 21 °C	65 – 69 °F	Somewhat uncomfortable for most people at upper edge	44% – 52%
16 – 18 °C	60 – 64 °F	OK for most, but all perceive the humidity at upper edge	37% – 46%
13 – 16 °C	55 – 59 °F	Comfortable	31% – 41%
10 – 12 °C	50 – 54 °F	Very comfortable	31% – 37%
< 10 °C	< 49 °F	A bit dry for some	30%

Humid heat: The heat capacity (specific heat) of humid air, expressed per unit mass of dry air in the mixture. The engineering symbol is c_S. Humid volume is defined as the total volume in cubic feet of lb of air and its accompanying water vapour. All this parameters can be determined by different methods like dew point method, psychrometric method, electric hygrometer method, gravimetric method, etc. In psychrometric method we use two thermometers; one dry bulb and the other wet bulb thermometer both are fixed to a frame. The wet bulb thermometer is in constant contact with water in order to achieve a temporary equilibrium with water at that temperature. The humidity can be determined with the help of humidity chart once the dry bulb and the wet bulb readings are obtained. The corresponding dew point, humid heat, humid volume and relative humidity can also be determined from the humidity chart.

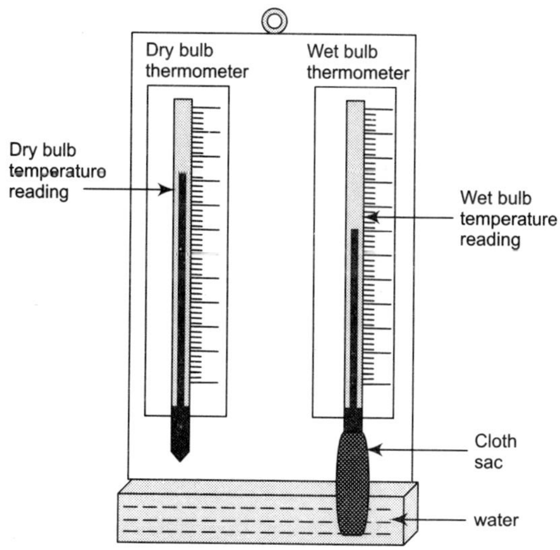

Fig. 25.1: Arrangement for the determination of humidity.

PROCEDURE

1. Arrange the wet bulb and dry bulb thermometers. One thermometer is exposed to the air (dry bulb thermometer) and the bulb of the second is covered with a cloth sac. (wet bulb thermometer)
2. Place water in the plastic container, and the cloth sack is dipped in to it. So that by capillary action the water will reach the bulb of the wet bulb thermometer and wet the bulb.
3. Carefully watch the readings in both the thermometers.
4. The reading in the wet bulb thermometer decreases as it is in direct contact with moisture content.
5. After some time the reading in the wet bulb thermometer reaches an equilibrium constant temperature. Note down this reading.
6. Simultaneously note down the dry bulb temperature also.
7. Take the 2 more readings after a time interval of 5 min.
8. Record the readings and find out the averages.

Determination of Humidity with the Help of Humidity Chart

1. Mark the dry bulb temperature and the wet bulb temperature on the humidity chart as A and B.
2. Draw straight line to the saturation curve parallel to the humidity axis. Line (I) and line (II).
3. Mark the point C where the line (II) meets the saturation curve.
4. From the point C draw a line (III) parallel to the adiabatic lines to the humidity axis.
5. Mark the point where the line (III) meets the line (II) as D.
6. Draw a straight line from the point D to the humidity axis parallel to the temperature axis. (Line IV)
7. Mark the point E where the line (IV) meets the humidity axis.
8. Note down the corresponding humidity at E. This is the humidity of the air at that temperature.
9. Extent the line (IV) to the saturation curve and mark the point F where it meets the saturation curve.

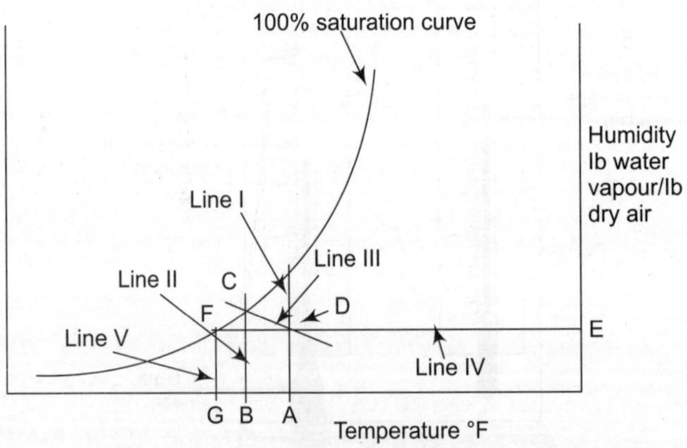

Fig. 25.2: Determination of humidity and dew point using humidity chart.

10. Draw a straight line V from the point F to the temperature axis parallel to the humidity axis.
11. Mark the point G where line touches the temperature axis.
12. This is the corresponding dew point temperature.
13. The experiment is repeated in different test conditions like outside the lab, air conditioned room etc.

Determination of Humid Heat, Humid Volume, Relative Humidity

14. Humid heat and humid volume can be interpreted from the corresponding line shown in the humidity chart.
15. Relative humidity can be determined by drawing a curve from point D parallel to the adjacent % humidity curves.

OBSERVATIONS AND CALCULATIONS

Table 25.1: Determination of Humidity and Dew Point

Sl No	*Condition*	*Dry bulb temperature(°C)*	*Dry bulb temperature(°F)*	*Wet bulb temperature(°C)*	*Wet bulb temperature(°F)*	*Humidity Lb water vapor/lb dry air*	*Dew point (°F)*	*Relative humidity (%)*	*Humid heat (BTU/°F/lb dry air)*	*Humid volume(cu.ft/lb dry air)*
1	Inside the lab									
2	Outside the lab									
3	Air cond. room									

REPORT

The humidity, Dew point, Relative humidity, humid heat and Humid volume of different conditions of atmosphere are studied.

EXPERIMENT 26

DETERMINATION OF HUMIDITY—DEW POINT METHOD

AIM

To determine the dew point of the air.

To determine the humidity of the air by dew point method.

To determine the wet bulb temperature at different condition of atmosphere.

To determine the dew point, relative humidity, humid heat, and humid volume.

To determine the level of human comfort of the atmosphere.

REQUIREMENTS

Psychrometric chart	(1)
Beaker	(3)
Thermometer	(3)
Burette stand	(3)
Glass rod	(3)
Water	
Ice	

PRINCIPLE

The dew point, or more precisely the dew point temperature, is the temperature at which the liquid and gaseous phases of a material present in a gas, such as water in air, are in equilibrium at a given gas pressure. In other words, the dew point temperature, or dew point, is the temperature at which the liquid water, or dew, evaporates at the same rate at which it condenses or It is the temperature to which the present air would have to be lowered to become saturated. Dew point is measured in degrees of temperature.

Dew point is a vital factor in convective heat transfer, for example combustion of fossil fuels and combustion engineering, drying of paper, cardboard, plastics, wood, tobacco, leather, printed goods, textiles and grain. The dew point temperature is most commonly observed in ambient air and is also called as the saturation temperature of water vapour in air. If you lower the temperature, dew will form as fog or condense on a cooler surface faster than it evaporates. This phenomenon is observed, for instance in a very moist air when the dew appears as fine water droplets suspended in air (fog) and on cool beverage containers in hot weather conditions.

Conversely, if the temperature increases above the saturation temperature, dew will evaporate faster than it condenses. Fog in air and dew on surfaces will disappear under such conditions. This is commonly seen when dew on the ground vanishes as the air warms during the day after a cool night.

The dew point temperature depends on the air temperature, since hotter air can hold more water vapour per unit volume than colder air. Likewise, the value of the dew point at a given air temperature is also a function of air pressure.

The dew point or saturation temperature is also described as the 100% relative humidity (RH) condition; when air holds 100% of the moisture that it possibly can. The moisture content of air can also be characterized in terms of the absolute humidity (AH) or humidity ratio, in terms of the weight of water vapour in the air per unit weight of dry air at the same temperature (pounds per pound or kilos per kilo). A high relative humidity indicates that the dew point is closer to the current air temperature. Relative humidity of 100% indicates the dew point is equal to the current temperature and the air is maximally saturated with water.

Humans tend to react with discomfort to a high dew point (*i.e.*, greater than 15 °C (59 °F)), as it interferes with the body's normal process of perspiration (producing sweat) to cool down. High relative humidity (which results in a high dew point) impedes the evaporation of sweat and reduces the effectiveness of evaporative cooling. As a result the body may overheat, resulting discomfort. Discomfort also exists when dealing with low dew points (*i.e.*, below – 30 °C (– 22.0 °F)). The dried air can cause skin to crack, become irritated more easily and will dry out the respiratory paths. (See the previous experiment). In this present study with the help of a simple experiment, the dew point of the air is determined. By using the psychrometric chart the other parameters like humidity, humid heat, relative humidity, humid volume can also be found out.

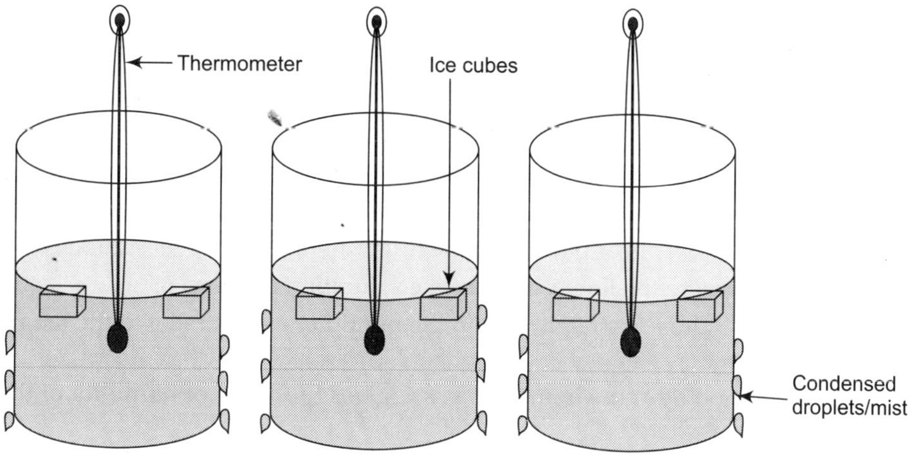

Fig. 26.1: Arrangement for the determination of dew point.

PROCEDURE

1. Arrange 3 beakers 250 ml shown in Fig. 26.1.
2. Add 150 ml of water into each of the beaker.
3. Arrange the thermometers in each of the beaker shown in Fig. 26.1.
4. Note down the temperature and wait to get constant temperature readings in all the beakers.
5. Add 4-5 pieces of ice cubes (small) in to the beaker.
6. Stir the beakers slowly in order to ensure a uniform cooling of the water.
7. Carefully watch the outer sides of the beakers and also the temperature readings.

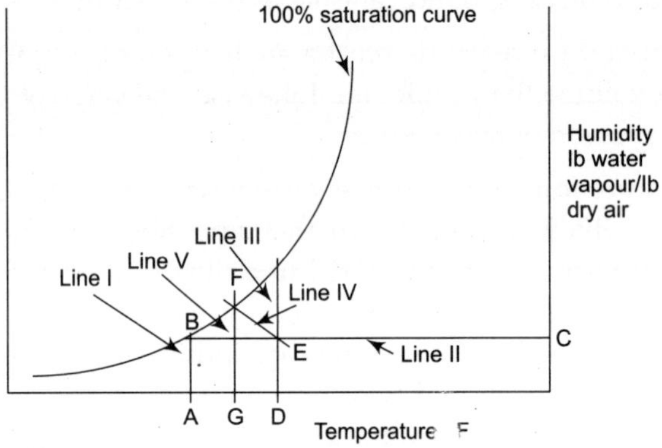

Fig. 26.2: Determination of humidity with the help of humidity chart.

8. Temperature of the water is noted when the first mist appears on the surface of the beakers.
9. This is the dew point of the air (saturation temperature).
10. Pour out the water and rearrange the beakers. Repeat the same experiments at least two times.
11. Find out the average of the 3 values.
12. The dry bulb temperature of the atmosphere is determined with the help of a normal dry bulb thermometer.
13. The experiment is repeated in different conditions like air conditioned room, and outside the lab in the open air.
14. Tabulate the readings.

To Find out the Humidity

1. The humidity can be determined with the help of the psychrometric chart.
2. Mark the dew point temperature (A) in the temperature axis of the psychrometric chart.
3. Draw a straight line from A parallel to the humidity axis to the saturation curve (line I)
4. Mark the point where the line A touches the saturation curve as B.
5. From this point B draw a straight line to the humidity axis parallel to the temperature axis. (line II)
6. The intersecting point in the humidity axis C gives the value of humidity of that particular condition of atmosphere.
7. Mark the corresponding dry bulb temperature in the temperature axis. (point D)

8. Draw a straight line to the saturation curve from the point D parallel to the humidity axis. (line III)
9. Mark the point E where the line II and line III meets.
10. Including point E draw a straight adiabatic line to the saturation curve parallel to the adjacent adiabatic lines (line IV.)
11. Mark the point F on the saturation curve where the line IV touch the saturation curve.
12. Draw another straight line from point F to the temperature axis parallel to the humidity axis (line V).
13. Note down the temperature on the temperature axis where the line meets the temperature axis (point G). This is the wet bulb temperature of that particular condition of atmosphere.

Determination of Relative Humidity, Humid Heat, Humid Volume

14. Humid heat and humid volume can be interpreted from the corresponding lines shown in the humidity chart.
15. Relative humidity can be determined by drawing a curve from point D parallel to the adjacent % humidity curves.

OBSERVATIONS AND CALCULATIONS

Table 26.1: Determination of Humidity

Sl No	*Test condition*	*Dew point temperature °C*				*Dry bulb temperature °C*				*Wet bulb temperature (°F)*	*Humidity (lb water vapor/lb dry air)*	*Humid heat (BTU/°F/lb dry air)*	*Humid Volume (cu.ft/lb dry air)*	*Relative humidity (%)*
		1	2	3	av	1	2	3	av					
1	Inside lab													
2	Outside lab													
3	Air cond. room													

REPORT

The dew point temperature is .... °F

Wet bulb temperature is .... °F

The humidity of the air is .... lb water vapour / lb of dry air

Relative humidity of the air .... %

Humid heat of the air .... BTU/°F/lb dry air

Humid volume ....cu.ft/lb dry air

The atmosphere of........dew point and...........% relative humidity is said to be ....

To human beings.

EXPERIMENT 27

DETERMINATION OF RELATIVE HUMIDITY BY WET BULB DEPRESSION METHOD

AIM

To determine the relative humidity by wet bulb depression method.

REQUIREMENTS

Wet bulb and dry bulb thermometers (mansion's type)	(1)
Humidity chart (Psychrometric chart)	(1)
Water	

PRINCIPLE

Water exists in the air at all temperatures. It is always in the vapour form. The quantity of the water vapour contained in the air occupies a very small part of air volume. The heat content of the air is the heat contained in the water vapour. Water vapour would condense when it comes in contact with any material at lower temperature. The quantity of water vapour present in the atmosphere is not constant as evaporation is continuously taking place from the surface of water in lakes, river and ocean and even from the human body. The quantity of the heat offered by the atmosphere is depends upon these evaporations.

Instead of the continuous evaporation taking place from the earth the air never become saturated. There is a maximum limit to the amount of water vapour that air can hold. The relative humidity of atmosphere is the ratio of mass of water vapour present in one cubic feet of atmospheric air to the mass of water vapour that air can hold at that temperature. It is expressed in percentage.

50% relative humidity mean that the air contains half the amount of water vapour that it can hold at that temperature. The comfort condition result from a desirable combination of temperature, humidity, air movement, and air cleanliness. If the temperature decreases the relative humidity increases. The temperature at which the relative humidity becomes 100% is called as the dew point. When the temperature of air is lowered below its dew point the water vapour suddenly starts condensing as snow, clouds, fog, or rain. Relative humidity can be determined by using relative humidity hydrometer, wet bulb and dry bulb thermometers, sling pshychrometer.

A mansion's type psychrometer is used in this experiment for determining the relative humidity by wet bulb depression method. It consists of two thermometers fixed on a wooden or plastic frame. One of the thermometers (wet bulb) is coved with cloth sack which is dipped in water taken in plastic container (See Fig. 27.1). The second thermometer (dry bulb) is free and exposed to the air. Unlike the sling psychrometer the assembly is fixed on the wall and not advised to swirl. Proper time should be given to reach the readings to equilibrium. After reaching equilibrium it is noticed that the reading in the wet bulb thermometer is always lower than the dry bulb thermometer. The amount of cooling at the wet bulb and the fall in the reading depends upon the rate of evaporation of water. It interns depends on the relative humidity of the air.

At low relative humidity rate of evaporation increases and more cooling is resulted, which will be responsible for a low wet bulb temperature. If the relative humidity is high evaporation rate decrease and a high wet bulb reading *i.e.*, not much difference from the dry bulb temperature.

The difference between the dry bulb temperature and wet bulb temperature is called as wet bulb depression. It is easy to predict the humidity conditions in terms of relative humidity by noting down the wet bulb depression. If the difference between the readings (dry bulb and wet bulb) increases relative humidity decreases. And if the difference between the readings decreases the relative humidity increases.

Wet bulb depression = Dry bulb temperature – Wet bulb temperature ... (Eq. 27.1)

With the help of the wet bulb depression chart provided it is possible to determine the exact percentage of relative humidity (Refer to Appendix VII).

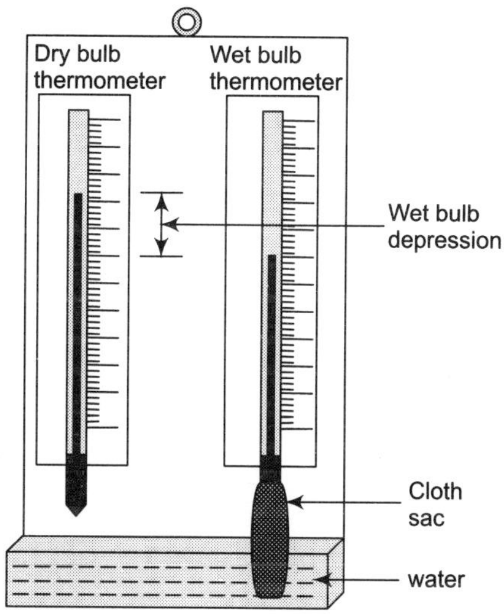

Fig. 27.1: Arrangement for the determination of wet bulb depression by dry bulb and wet bulb thermometers (Mansion's type).

PROCEDURE

1. Arrange the wet bulb and dry bulb thermometers. One thermometer is exposed to the air (dry bulb thermometer) and the bulb of the second is covered with a cloth sac. (wet bulb thermometer)

2. Place water in the plastic container, and the cloth sac is dipped in to it. So that by capillary action the water will reach the bulb of the wet bulb thermometer and wet the bulb.
3. Carefully watch the readings in both the thermometer.
4. The reading in the wet bulb thermometer is decreases as it is in direct contact with moisture content.
5. After some time the reading in the wet bulb thermometer reaches an equilibrium constant temperature and note down this reading.
6. Simultaneously note down the dry bulb temperature.
7. Take 2 more readings after a time interval of 5 min.
8. Record the values and find out the difference between the wet bulb and dry bulb temperatures.
9. This value is called as the wet bulb depression.
10. After calculating the wet bulb depression the relative humidity can be found directly from table provided. (Refer to appendix VII).
11. Locate the reading of dry bulb and wet bulb depression value and the intersection of two columns read the percentage of relative humidity.
12. The same experiment is repeated in the open air outside the lab and in an air conditioned room.

OBSERVATIONS AND CALCULATIONS

Table 27.1: Determination of Relative Humidity

Sl No	*Condition*	*Dry bulb temperature °C*				*DET (°F)*	*Wet point temperature °C*				*WBT (°F)*	*Wet bulb depression*	*RG (%)*
		1	2	3	Av		1	2	3	Av			
1	Inside lab												
2	Outside lab												
3	Air cond. room												

REPORT

The relative humidity of air at different conditions is determined. It is found that the relative humidity is in the order of ....> ....> ....

EXPERIMENT 28

DETERMINATION OF RELATIVE HUMIDITY BY SLING PSYCHROMETER

AIM

To determine the relative humidity by wet bulb depression method using sling psychrometer.

REQUIREMENTS

Sling psychrometer	(1)
Humidity chart (wet bulb depression chart)	(1)

PRINCIPLE

(Refer to previous experiment).

In this experiment relative humidity is determined by sling psychrometer. It consists of two thermometers fixed on a metallic frame (Refer to Fig. 28.1) which can be rotated by a handle. One of the thermometers (wet bulb) is covered with cloth sack which can be dipped in water (See Fig. 28.1). The second thermometer (dry bulb) is free and exposed to the air. The sling psychrometer is whirled for some seconds until the reading in the wet bulb thermometer reaches equilibrium. After reaching equilibrium it is noticed that the reading in the wet bulb thermometer is always lower than the dry bulb thermometer. The amount of cooling at the wet bulb and the fall in the reading depends upon the rate of evaporation. The rate of evaporation interns depends on the relative humidity of the air.

At low relative humidity rate of evaporation increases and more cooling is resulted, which will be responsible for a low wet bulb temperature. If the relative humidity is high evaporation rate decrease and a high wet bulb reading exist *i.e.*, not much difference from the dry bulb temperature.

It is easy to predict the humidity conditions in terms of relative humidity by noting down the wet bulb depression. If the difference between the readings (dry bulb and wet bulb) increases relative humidity decreases. And if the difference between the readings decreases the relative humidity increases.

Wet bulb depression = Dry bulb temperature – Wet bulb temperature ... (Eq. 28.1)

With the help of the wet bulb depression chart provided it is possible to determine the exact percentage of relative humidity. (Refer to appendix VII)

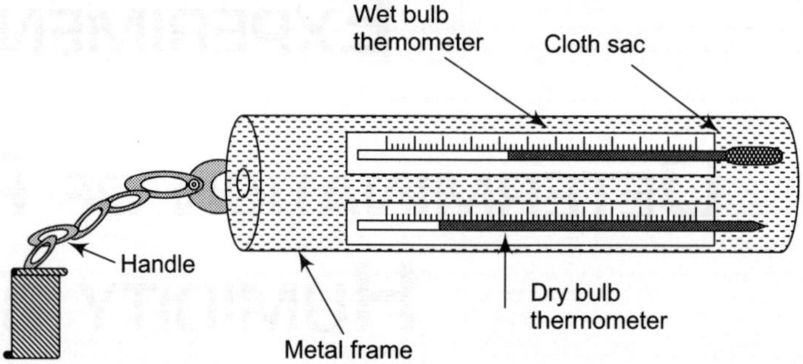

Fig. 28.1: Determination of wet bulb depression by sling psychrometer.

PROCEDURE

1. Arrange the Sling psychrometer. One thermometer is exposed to the air (dry bulb thermometer) and the bulb of the second is covered with a cloth sac. (Wet bulb thermometer)
2. The cloth sac is dipped in water so that by capillary action the water will reach the bulb of the wet bulb thermometer.
3. The sling psychrometer is rotated on its handle for 1-2 min.
4. The reading in the wet bulb thermometer is decreases as it is in direct contact with moisture content.
5. The readings in both the thermometers will reach an equilibrium constant temperature and note down this readings.
6. Take the 2 more readings after a time interval of 5 min.
7. Record the values and find out the difference between the wet bulb and dry bulb temperatures.
8. This value is called as the wet bulb depression.
9. After calculating the wet bulb depression the relative humidity can be found directly from table provided.

OBSERVATIONS AND CALCULATIONS

Dry Bulb temperature =.............

Wet bulb temperature =............

Wet bulb depression = Dry bulb temperature – Wet bulb temperature

= ....

Table 28.1: Determination of Relative Humidity by Wet Bulb Depression Method

Sl No	*Condition*	*Dry bulb temperature (°C)*				*DBT (°F)*	*Wet bulb temperature (°C)*				*WBT (°F)*	*Wet bulb depression*	*RH (%)*
		1	**2**	**3**	*Av*		**1**	**2**	**3**	*Av*			
1	Inside the lab												
2	Outside the lab												
3	Air cond. room												

REPORT

The relative humidity of air at different conditions is determined by sling psychrometer.

EXPERIMENT 29

DRYING—DETERMINATION RATE OF DRYING

AIM

To perform the drying of the given sample.
To demonstrate the behaviour of the drug during drying.
To determine the rate of drying.

REQUIREMENTS

Petri dish	(1)
Motar & Pestle	(1)
Glass rod	(1)
Hot air oven	(1)
Weighing balance	(1)
Beaker	(1)
$CaCO_3$	

PRINCIPLE

Drying is the process of removal of water or other moisture or other liquid from a substance with the aid of heat. In all the substances the moisture will be present in three forms; bound moisture, unbound moisture, free moisture. Absorption or desorption of moisture is continued till the equilibrium moisture content is reached. EMC is the amount water present in the solid which is in equilibrium with atmospheric moisture. Normal drying process is proceeded until the EMC is reached. Any attempt to remove moisture content below the EMC is the wastage of time and energy.

The rate of drying can be determined by performing a simple experiment with the help of a Petri dish and hot air oven. Prepared slurry in a Petri dish (previously weighed) is dried in a hot air oven and the weight lost with time is recorded. The water present in the slurry will be moved to the surface by capillary action as well as diffusion. At the time it reaches the surface of the slurry evaporation starts because of the temperature gradient set in the atmosphere of the dryer. This process will continue till the equilibrium moisture content (EMC) is reached.

The rate of drying during the process can be calculated with the help of the equation,

$$\textbf{Rate of drying} = \frac{\textbf{Weight of moisture removed}}{\textbf{Weight of the drug (dry)} \times \textbf{Time} \times \textbf{Surface area exposed}} \quad \text{... (Eq. 29.1)}$$

The behaviour of the given sample during drying can be explained by plotting various graphs as given below. The critical moisture content (CMC), constant rate period, first falling rate period, second falling rate period, second critical point, second falling rate period, and EMC can be interpreted from the graphs.

PROCEDURE

1. Accurately weigh 5 gm of $CaCO_3$.
2. Place in a mortar and add sufficient amount of water (10 ml) to make slurry.
3. Weigh a Petri dish (W_1) and its diameter is noted.
4. Uniformly spread the prepared slurry in the Petri dish.
5. Weigh the Petri dish with the slurry and weight is noted as W_2.
6. Set the hot air oven at 60°C.
7. Place the Petri dish in the oven with the lid partially open.
8. Take out the dish at intervals of 10, 20, 30, 45, 60, 75, and 90 and weight of the petridish + slurry is noted.
9. Proceed till the constant weight is obtained.
10. Record the data obtained.
11. Find out the moisture removed at each time interval.
12. Calculate the corresponding rate of drying.
13. Draw the graphs while taking the appropriate values at the coordinates.
14. Mark the critical moisture content, first falling rate period, second falling rate period etc as shown in the figures.

OBSERVATIONS AND CALCULATIONS

Diameter of the Petri dish = ————————cm

Area of the Petri dish = ————————cm^2

Weight of the Petri dish (W_1) = ————————gm

Weight of Petri dish + slurry (W_2) = ————————gm

MODEL CALCULATIONS

Weight of the slurry = W_2 - W_1 ......gm

FMC = Weight of the slurry – 5 = ...gm

Weight of the moisture removed = A – B = ...gm

(A is the weight of the slurry at 0 min and B is the weight of the slurry at 10 min)

$$\text{Rate of drying} = \frac{\text{Weight of moisture removed}}{5 \times 10 \times \text{Area of the Petri dish in cm}^2}$$

$$= \ldots C \ldots\ldots g/g.\ cm^2.\ 10\ min$$

$$\text{Rate of drying} = C \times 60/10$$

$$= \ldots\ldots\ldots g/g\ .\ cm^2.\ hr$$

Table 29.1: Determination of Rate of Drying

Sl No	*Drying time (min)*	*Weight of slurry (gm)*	*FMC (gm)*	*Weight of Moisture removed (gm)*	*Rate of drying (gm/g.cm².10 min)*	*Rate of drying (Gm/gm.cm² hr)*
1	0					
2	10					
3	20					
4	30					
5	45					
6	60					
7	75					
8	90					
9	120					

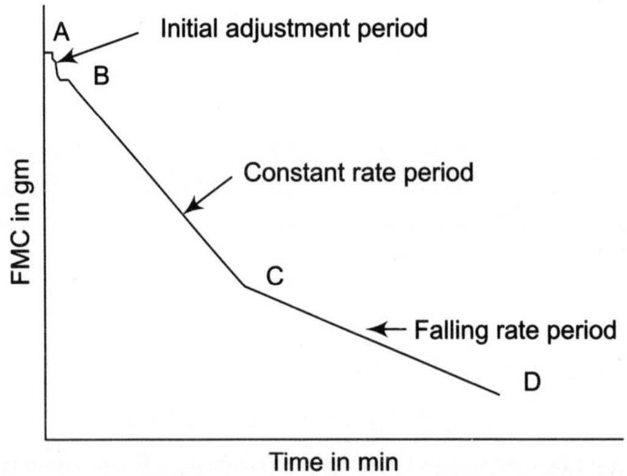

Fig. 29.1: Graph of moisture content Vs Time.

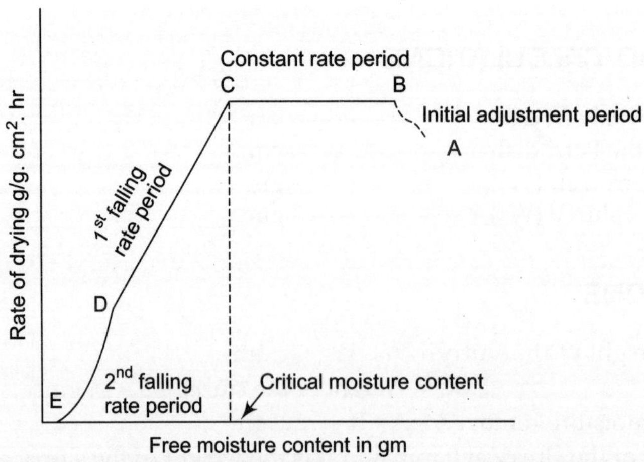

Fig. 29.2: Drying rate curve (Rate of drying Vs Moisture content).

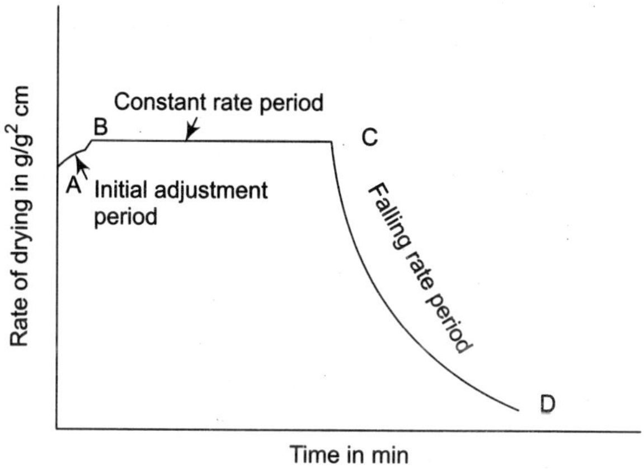

Fig. 29.3: Drying rate curve (Rate of drying Vs Time).

REPORT

The drying of $CaCO_3$ slurry is performed and the drying pattern is shown in terms of graphs.

The rate of drying of the $CaCO_3$ slurry = ... g/g.cm^2.hr

The critical moisture content = ... gm

EMC = ... gm

EXPERIMENT 30

DRYING—COMPARISON OF THE DRYING PATTERN OF SOLIDS

AIM

To demonstrate the drying pattern of different solids.

To categorize the solids depending upon their behaviour during drying.

REQUIREMENTS

Petri dish	(2)
Motar & pestle	(1)
Glass rod	(2)
Hot air oven	(1)
Weighing balance	(1)
Beaker	(1)
$CaCO_3$	

PRINCIPLE

Drying is the process of removal of water or moisture or other liquid from a substance with the aid of heat. In all the substance the moisture will be present in three forms; bound moisture, unbound moisture, free moisture. Solids drugs will show different pattern of drying. For example, solids drugs are roughly classified as:

1. Granular or crystalline
2. Amorphous solids.

These two categories of solids differ greatly in the drying behaviours. In Granular solids the water is held in shallow open surface pores and interstitial space between particles that are easily accessible to the surface *e.g.*, calcium sulphate, zinc oxide, magnesium oxide. The moisture in crystalline solids is lost with little hindrance by gravitational or capillary force. If you are plotting drying curve, the constant rate period is the major portion of the drying curve, the falling rate period is much shorter and EMC is close to zero. But in case of Amorphous solids the moisture is an integral part of the molecular structure as well as being physically entrapped in fine capillaries and small interior pores *e.g.*, Starch, casein, yeast, insulin. During drying in this type of solids the movement of water vapour is slow. Drying pattern shows a short constant rate period and long critical rate period. EMC is seen high as most of the water remains intimately associated within the molecular interstitial space of the surface.

In this present study, we are attempting to study the drying patterns of two entirely different samples starch and calcium carbonate.

PROCEDURE

1. Accurately weigh 5 gm of $CaCO_3$.
2. Place in a mortar and add sufficient amount of water (10 ml) to make slurry.
3. Weigh a Petri dish (W_1) and its diameter is noted.
4. Spread the prepared slurry in the Petri dish uniformly.
5. Weigh the Petri dish with the slurry and weight is noted as W_2.
6. Set the hot air oven at 60 °C.
7. Place the Petri dish in the oven with the lid partially open.
8. Take out the dish at interval of 10, 20, 30, 45, 60, 75, 90 and the weight of the dish along with the slurry is noted.
9. This is continued till a constant weight is obtained.
10. Record the data obtained.
11. Find out the moisture removed at each time interval.
12. Calculate the corresponding rate of drying.
13. Draw the graphs while taking the appropriate values at the coordinates.
14. The same procedure is repeated for 5 gm of starch.
15. Compare the results obtained.
16. The drying pattern of both the slurry can be studied by plotting the graphs. (Refer to previous experiment)

OBSERVATIONS AND CALCULATIONS

Diameter of the Petri dish = ——————— cm

Area of the Petri dish = ——————— cm^2

Weight of the Petri dish (W_1) = ——————— gm

Weight of Petri dish + slurry (W_2) = ——————— gm

Weight of the slurry $W_2 - W_1$ = ——————— gm

Table 30.1: Determination of Rate of Drying for Calcium Carbonate

Sl No	Drying time (min)	Weight of slurry (gm)	FMC	Weight of Moisture removed (gm)	Rate of drying (gm/g.cm^2.10min)	Rate of drying (gm/gm.cm^2.hr)	EMC
1	0						
2	10						
3	20						
4	30						
5	45						
6	60						
7	75						
8	90						
9	120						

Table 30.2: Determination of Rate of Drying for Starch

Sl No	Drying time in min	Weight of slurry (gm)	FMC	Weight of Moisture removed (gm)	Rate of drying (gm/ g.cm^2.10 min)	Rate of drying (gm/gm.cm^2.hr)	EMC
1	0						
2	10						
3	20						
4	30						
5	45						
6	60						
7	75						
8	90						
9	120						

MODEL GRAPHS

(Refer to previous experiment)

REPORT

The drying patterns of two different drugs are determined experimentally.

The Critical moisture content of the calcium carbonate is = ... gm

The Critical moisture content of starch = ... gm

The constant rate period last for (calcium carbonate) = ... min

The constant rate period lasts for (starch) = ... min

EMC of starch = ... gm

EMC of calcium carbonate = ... gm

Hence ....is considered as crystalline nature and .... is considered as amorphous nature.

EXPERIMENT 31

DETERMINATION OF EQUILIBRIUM MOISTURE CONTENT

AIM

To determine the equilibrium moisture content of the given sample of material at different humidity conditions.

REQUIREMENTS

Desiccators	(9)
Bottles	(18)
Beakers	(9)
Glass rods	(4)
Weighing balance	(1)
Weight box	(1)

Salts for preparing saturated solutions (Lithium bromide, Potassium acetate, Magnesium chloride, Potassium carbonate, Magnesium nitrate, Sodium nitrate, Sodium chloride, Potassium bromide, Potassium nitrate).

PRINCIPLE

Exposure to air at a definite temperature and humidity will cause a material to lose or gain moisture until an equilibrium content is reached. The portion of water that cannot be removed by the air because of the humidity of the air is called as equilibrium moisture content (EMC). Suppose that a wet mass is brought in contact with a stream of air having certain temperature and humidity (both are constant) the moisture from the wet mass starts to evaporate and after a sufficiently long time of exposure an equilibrium is reached and the material will retain a certain amount of moisture content which will be unchanged with the exposure of the same air. This moisture content is called as EMC. Even if the material is reduced to a bone dry state an exposure to the atmosphere will produce the regaining of the moisture untill the EMC is reached will be a regain of the moisture till the EMC is reached. EMC value will vary depending on the temperature and humidity and the properties of the material. The moisture content is expressed as percentage moisture content, dry basis (number of parts of weight of moisture per hundred parts of bone dry solid). Determination of EMC is helpful to avoid the over drying of the products. Some materials are having high EMC content at normal atmospheric temperature and humidity. By knowing the values of EMC of each raw materials used in the preparation a better optimization of process can be achieved.

PROCEDURE

1. Take 9 desiccators and clean properly.
2. Prepare the saturated solutions of different salts. (Refer to Appendix III)
3. Take 18 glass bottles of same capacity and weigh it. (15 ml).
4. Mark all the bottles. (*i.e.*, LC1 & LC2 for placing in the desiccators containing lithium chloride, PA1 & PA2 for potassium acetate etc)
5. Prepare 18 packets containing 500 mg of same substance (kaolin)
6. Add the contents to each bottle carefully. (500 mg in each bottle)
7. Weigh all the bottles individually and record the weights.
8. A liberal amount of saturated salt solution (with excess crystals) is added in to the well of the desiccators.
9. Label the desiccators according to the saturated solution it contains. (If lithium chloride is the saturated solution, mark the desiccators as LC).
10. Place two bottles containing samples (open) in each desiccators and properly mark the bottles (care should be taken to place the bottles labeled with LC1 & LC2 in the desiccators containing lithium chloride solution and PA1 & PA2 in the desiccators containing potassium acetate and so on).
11. Samples are placed in an upright position.
12. Securely close all the desiccators and store all nine samples (securely closed individual desiccators, each have a different moisture atmosphere) for 7 days at controlled room temperature of 25 °C ± 2 °C
13. After 7 days all the bottles are carefully removed and individually weighed.
14. The increase or decrease of weight is recorded accordingly.
15. Find out the percentage of EMC.

OBSERVATIONS AND CALCULATIONS

Table 31.1: Determination of EMC

Sl No	*Humidity condition %RH*	*Bottle Number*	*Weight of empty bottle gm* W_1	*Weight of empty bottle with sample* W_2	*Weight of sample + bottle after 7 days (gm)* W_3	*Weight of sample after 7 days(gm)* $W_4 = W_3 - W_1$	*Moisture absorbed gm* $W_5 = W_4 - W_2$	*% of moisture absorbed* $(W_5/0.5) \times 100$	*Average % of moisture absorbed*
	11	LC1							
		LC2							
	23	PA1							
		PA2							
	33	MC1							
		MC2							
	43	PC1							
		PC2							
	52	MN1							
		MN2							
	64	SN1							
		SN2							
	75	SC1							
		SC2							
	83	PB1							
		PB2							
	93	PN1							
		PN2							

REPORT

The EMC at different humidity conditions are determined. The EMC .... with increase in RH

EXPERIMENT 32

FILTRATION—EFFECT OF FILTER AIDS ON RATE OF FILTRATION

AIM

To study the influence of filter aids on the rate of filtration.

REQUIREMENTS

Beaker	(3)	Bentonite
Measuring cylinder	(3)	Kieselghur
Glass rod	(3)	Talc
Funnel	(3)	Calcium carbonate
Motor and pestle	(3)	
Burette stands	(3)	
Filter paper		

PRINCIPLE

Filtration is the process of separation of fluid from solid with the help a filter media. In the case of separation of solid from liquid, the solid separated out is called as filter cake and the liquid coming out from the filter media is called as filtrate. If slurry subjected to filtration containing slimy compressible material, the filter media is soon clogged and it may be necessary to incorporate filter aids. A filter aid is a solid material, finely divided but consisting of hard, strong, particle that are incompressible. The most common filter aid used is diatomaceous earth or kieselghur. Filter aids function by building up a porous rigid structure, which can retain the solid particle and allow the fluid to pass through without clogging the filter cloth.

Filter aids can be used in two ways. It can be added to the suspension and mixed with solution to be filtered. Or after making solution of filter aid in a suitable liquid, poured over the filter media in order to produce a pre coat. Examples of other filter aids are talc, Bentonite etc. The present experiment demonstrates how the addition of filter aids affects the rate of filtration.

PROCEDURE

1. Prepare a slurry of $CaCO_3$ (5%) in 300 ml of water.
2. Equally divide in to three parts and taken in to 3 separate beakers and mark it as I, II, III
3. Weigh 500 mg of bentonite and kieselghur separately.
4. Arrange 3 filtration assemble. (Measuring cylinder, Funnel, burette stand, Filter paper)
5. Mark it as I, II, III.
6. No upstream or downstream pressure is applied during filtration.
7. Pour the slurry from the beaker I to the assembly I (stop adding when the slurry fills in the funnel ¾th. During each addition shake the beaker thoroughly in order to avoid the sedimentation of the particles).

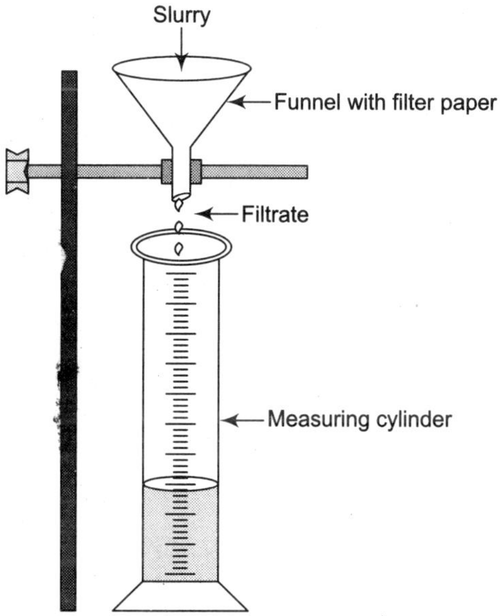

Fig. 32.1: Assemble for filtration.

8. Find out the amount of filtrate collected at different time interval of 5, 10, 15, 20, 25, 30, 45 min.
9. To the second beaker 500 mg of bentonite is added and stirred well.
10. Now add slowly the slurry of $CaCO_3$ over the filter assemble II.
11. Note down the amount of filtrate collected at different time interval of 5, 10, 15, 20, 25, 30, 45 min.
12. The steps 9-12 is repeated with the filter aid kieselghur using the III set of filtration assembly and the slurry from the beaker III.
13. The data are tabulated.

OBSERVATIONS AND CALCULATIONS

Filter volume = ...ml

Time = ...

Rate of filtration = ...ml/min

Table 32.1: Determination of Rate of Filtration

Sl No	*Time (min)*	*Filtrate volume (ml)*			*Rate of filtration ml/min*		
		I (Without filter aid)	II (Bentonite)	III (Kieselghur)	I	II	III
1	5						
2	10						
3	15						
4	20						
5	25						
6	30						
7	45						
8	60						
9	75						

REPORT

The rate of filtration .... with the use of filter aids. The order of increase in rate of filtration using ....>.........>............

EXPERIMENT 33

EFFECT OF FILTER AID CONCENTRATION ON RATE OF FILTRATION

AIM

To determine the ideal concentration of filter aid for the given suspension of $CaCO_3$.

REQUIREMENTS

Beaker (1000 ml)	(1)
Beakers (500 ml)	(6)
Buchner funnels	(6)
Glass rod	(6)
Weighing balance	(1)
Filter papers	
keiselguhr	

PRINCIPLE

Filtration is the process of separation of fluid from solid with the help a filter media. In the case of separation of solid from liquid, the solid separated out is called as filter cake and the liquid coming out from the filter media is called as filtrate. As the filtration proceeds the filter cake formed over the filter media starts to offer a resistance to the flow of fluid through it. This will affect the total rate of filtration. The solids can either penetrate the void space of the filter media or it can produce a bridging over the openings or produce an impermeable cake. This increases the resistance to flow.

Filter aids are a special type of filter medium. Filter aid acts by reducing the resistance offered by the filter cake. The filter aid forms a fine surface deposit that screen out all solids, preventing them from contacting and plugging the supporting filter media. The quantity of the filter aid greatly influences the filtration rate. If too little filter aids are used, the resistance offered by the filter cake is greater than if no filter aid is used, because of the added thickness to the cake. On the other hand, if high amount of filter aid is added the filter aid merely aid to the thickness of the cake without providing the cake porosity.

In this experiment, suspensions of $CaCO_3$ containing increasing concentrations of filter aid are prepared and the effect of concentration of filter aid on the filtration rate is noted. Graph of concentration of filter aid in % on X axis and flow rate on Y axis is plotted to show the effect of filter aid concentration on filtration.

PROCEDURE

1. Weigh accurately 20 gm of $CaCO_3$ and add into a mortar.
2. Add sufficient amount of water to produce fine slurry. (Measure the water)
3. Transfer the slurry into a 1000 ml beaker.
4. Wash the motor with 50 ml of water and add this also to the beaker.
5. Add water to the beaker to produce 800 ml.
6. Stir the suspension in order to avoid the sedimentation.
7. Transfer equal quantity of suspension to 5 small beakers (100 ml each) and mark it 1-V.
8. Arrange five filtration assemblies as shown in Fig. 33.1 and number it 1-V.
9. Add 100 mg (0.1%) of the keiselguhr into the first beaker and stir to ensure uniform distribution.

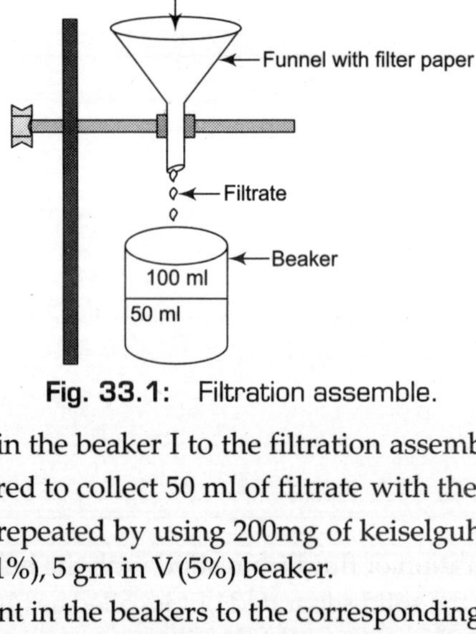

Fig. 33.1: Filtration assemble.

10. Pour the total contents in the beaker I to the filtration assemble I.
11. Find out the time required to collect 50 ml of filtrate with the help of a stopwatch.
12. The same procedure is repeated by using 200mg of keiselguhr in the II (0.2%), 500 mg in III (0.5%), and 1 gm in IV (1%), 5 gm in V (5%) beaker.
13. Pour the contents present in the beakers to the corresponding filtration assemble (ie content of the II beaker to the Filtration assemble II, and content of beaker III to filtration assemble III and so on).
14. Repeat the steps 11-12.
15. A blank determination is also carried out without adding the filter aid.
16. Record the observations.
17. Find out the rate filtration.
18. Plot a graph of % of filter aid in X axis and rate of filtration on Y axis.
19. Report thc findings.

OBSERVATIONS AND CALCULATIONS

Concentration of the suspension = 20/800 × 100

= ......%.......

Concentration of filter aid

1st beaker = 0.1/100 × 100

= ...0.1.......%

2nd beaker = 0.2/100 × 100

= ... 0.2.........%

3rd beaker = 0.5/100 × 100

= ......0.5.... %

4th beaker = 1/100 × 100

= ...1....... %

5th beaker = 5/100 × 100

= ...5..........%

Rate of filtration = .../.......

Table 33.1: Determination of Effect of Filter Aid Concentration of Rate of Filtration

Sl No	*Concentration of filter aid (%)*	*Time to collect 50 ml of filtrate (min)*	*Time to collect 50 ml of filtrate (sec)*	*Rate of filtration (ml/sec)*
1	0.1			
2	0.2			
3	0.5			
4	1			
5	5			
6	0			

REPORT

The ideal concentration of keiselguhr for the filtration of $CaCO_3$ suspension is found to be ....%.

EXPERIMENT 34

EVALUATION OF FILTER MEDIA

AIM

To evaluate the given filter media.

REQUIREMENTS

Filter media (sand, cotton, cotton cloth, polyester cloth, filter paper, parchment paper, membrane filter)

Beakers	(7)
Funnels	(7)
Glass rod	(3)
Test tubes	(7)
Test tubes stand	(1)
Sintered glass funnels	(1)
Conical flask	(1)
Water	
$CaCO_3$	

PRINCIPLE

Filtration is defined as a process in which the particles are separated from a liquid by passing the liquid through a permeable material. According to the mechanism, the filtration can be classified in to two types, surface filtration and depth filtration.

In surface filtration a screening action takes place by pores or holes. The solids are got deposited over filter medium. Filter media are available in different forms. Sand, filter cloth, felts, bonded fabrics, kraft papers, membrane filters, surface type cartridges, are some of the filter media commonly used. During the filtration, depending upon the pore size, particles present in the slurry will either retain over the media or pass through the filter media.

In this experiment a number of filter media are tested for their efficiency to filter slurry. The filtrate is analyzed with the help of turbidimeter to measure the turbidity produced by the presence of infiltrated particles. The turbidity can be measured in terms of % transmittance or absorbance of light through the sample. The more turbid the filtrate, the less light will pass through it and lesser the turbidity greater will be the efficiency of filter media.

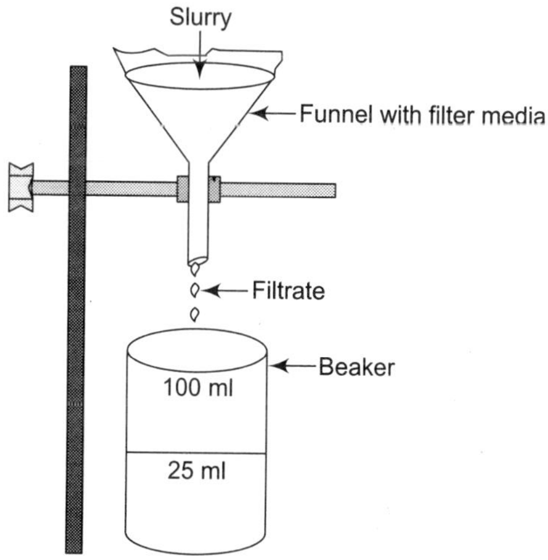

Fig. 34.1: Filtration assemble

PROCEDURE

Filtration of the Slurry

1. Arrange 7 filtration assemble as shown in Fig. 34.1.
2. Keep different filtration media to the funnels.
3. Prepare the slurry of $CaCO_3$ (10%).
4. Add 50 ml slurry to each of the filter media. (Stir the slurry before each addition)
5. Discard the first few drops.
6. Wait till 10 ml of filtrate is collected.
7. Take the filtrates separately into test tubes and mark it accordingly.
8. Analyze the samples with the help of spectrophotometer.

   Note: For membrane filter use a sintered glass funnel and conical flask.

Analysis of the Filtrate by Turbidimeter.

9. Turn the instrument on.
10. Allow the instrument to warm up for 10-15 min.
11. Set the wave length.
12. Set percentage transmittance to 0% by turning the knob.
13. Clean the cuvette and take pure water in that. (Blank)
14. Wipe the outer walls of the cuvette with a tissue paper and insert in to the tube holder.

15. Close the cover.
16. Set the reading of 100% transmittance while turning the knob.
17. Remove the blank from the cuvette and place the first sample.
18. Wipe off the outer walls of the cuvette with tissue paper and insert into the tube holder.
19. Close the cover.
20. Note down the % of transmittance.
21. The same procedure is repeated for all other samples. (Steps 11-18)
22. The findings are tabulated and reported.

Table 34.1: Determination of Percentage Transmittance

Sl No	*Samples*	*% transmittance*
1		
2		
3		
4		
5		
6		
7		
8	Blank	

REPORT

The given filter media are evaluated.

By keeping the measurement of % transmittance as a tool, the order of filter media giving good separation is ....>........> ....> ....>........> ......>......

EXPERIMENT 35

CENTRIFUGATION—DETERMINATION OF THE YIELD STRESS OF CENTRIFUGED CAKE

AIM

To determine the yield stress of centrifuged cake.

To demonstrate the buoyancy effect with help of rods of different density and diameter.

REQUIREMENTS

Centrifuge	(1)
Glass, aluminum, wood rods	
(Different diameter and length)	(3)
Burette stands	(3)
Cotton	
$CaCO_3$	

PRINCIPLE

Buoyancy is the phenomenon (discovered by Archimedes) that an object less dense than a fluid will float in the fluid. More generally, Archimedes' principle states that a fluid will exert an upward force on an object immersed in it equal to the weight of the fluid displaced by the object. By definition, sedimentation force (F_S) is the downward force of weight due to linear acceleration (mg) minus buoyancy (F_B) minus frictional forces opposing downward motion (F_F).

The objective of the spin tube test is to check the settling ability of solids in slurry under centrifugation. A small or equal amount of feed is introduced into a number of test tubes with a volume of 15-50 ml. the samples are centrifuged at a certain rpm for a period of time *t*. The supernatant liquid is decanted off. The integrity more precisely the yield stress of the cake can be determined by finding out the amount of penetration of a rod into the cake under its weight. The yield stress of the centrifuged cake can be determined by

$$\text{Yield stress} = \frac{(\rho_r L - \rho_L)^{1/2}}{h} \times g\, r_d \qquad \text{... (Eq. 35.1)}$$

where,

r_d = Radius of the rod (cm)
L = Length of the rod (cm)
h = Length of penetration of the rod (cm)
ρ_L = Density of the cake (g/cm^3)
g = Acceleration due to gravity (cm/sec^2)
ρ_r = Density of the rod (g/cm^3)

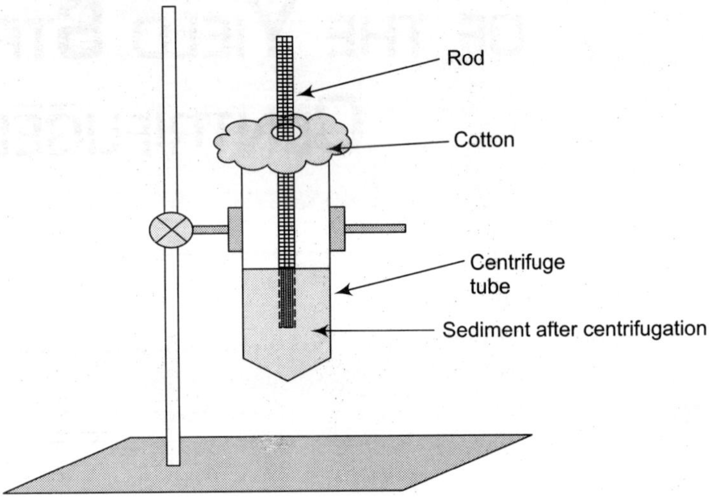

Fig. 35.1: Arrangement for determination of yield stress.

PROCEDURE

1. Prepare a known concentration of suspension of any drug ($CaCO_3$) by simply trituration method.
2. Fill equal amount of the suspension in the centrifugal tubes. (The volume can be selected according to the capacity of the tube (normally 15-50 ml)
3. Arrange the samples properly in the centrifuge to avoid the accidental breakage during operation.
4. Allow to operate the centrifuge at 1000 RPM for 10 min.
5. Take out all the tubes carefully.
6. Decant the supernatant liquid without disturbing the cake.
7. Clamp the tubes on a burette stands shown in Fig. 35.1.
8. Find out the diameter and length of the rods provided.
9. Close the mouth of the tubes with cotton.
10. Make a hole at the centre of the cotton so as to pass the rods freely through the cotton (Fig. 35.1).
11. Carefully place the glass rod in the tube I. (One in each tube)

12. Care should be taken not to give pressure on the rod while inserting.
13. The penetration of the rod to the cake will depends upon the upward force exhibited by the wet cake.
14. Wait for 1-2 min and remove the cotton from the mouth. Carefully remove the rod from the tube.
15. Immediately measure the length of wet portion of the rod.
16. Place the aluminum rod in the second assemble of centrifugal tube.
17. Steps 12-15 are repeated.
18. The experiment is repeated with other rods using fresh assemble.
19. Tabulate the data obtained.
20. The same procedure can be repeated for different RPM and time of centrifugation and different length and diameter of rods.
21. Find out the yield stress for all the conditions.

OBSERVATIONS AND CALCULATIONS

Table 35.1: Determination of Yield Stress

Sl No	Rpm	Time(min)	*Nature of the material*											
			Wood				*Glass*				*Aluminum*			
			Length of rod(cm)	Radius of the rod (cm)	Length of penetration(cm)	Yield stress (cm^3/sec)	Length of rod (cm)	Radius of rod cm	Length of penetration (cm)	Yield stress (cm^3/sec)	Length of rod(cm)	Radius of rod cm	Length of penetration cm	Yield stress (cm^3/sec)

Weight of the $CaCO_3$ taken = ...gm

Volume of the wet cake occupied

after centrifugation = ...ml

Density of the cake (ρ_L) = ... gm/ml

$$\text{Yield stress} = \frac{(\rho_r L - \rho_L)^{1/2}}{h} \times g\, r_d$$

$$= \text{..............}\ cm^3/sec$$

REPORT

Yield stress using wood rod

(…diameter …..Length) = …………………….. cm^3/sec

(…diameter …..Length) = …………………….. cm^3/sec

Yield stress using glass rod

(…diameter …..Length) = …………………….. cm^3/sec

(…diameter …..Length) = …………………….. cm^3/sec

Yield stress using aluminum rod

(…diameter …..Length) = …………………….. cm^3/sec

(…diameter …..Length) = …………………….. cm^3/sec

EXPERIMENT 36

CENTRIFUGATION—DETERMINATION OF CENTRIFUGAL EFFECT

AIM

To determine the centrifugal effect during the separation of emulsion at different rpm.

REQUIREMENTS

Centrifuge	(1)
Mortar and pestle	(1)
Beaker	(1)
Measuring cylinder	(1)
Glass rod	(1)
Castor oil	
Acacia	
Water	

PRINCIPLE

Centrifugation is based on the well known principle (of centrifugal force) that an object which is rotated about a centre point at a constant radial distance from the point is acted upon by a force. The centrifugal force is measured in terms of the number of times the centrifugal force is greater than that of the gravitational force. During centrifugation the centrifugal effect acted on the body under rotation, is proportional to the speed of rotation and diameter of rotation. The centrifugal effect can be determined with the help of the equation,

$$\boxed{\mathbf{C.E = 2.012}\ n^2\ d} \quad \text{... (Eq. 36.1)}$$

where, n = Number/speed of rotation (s^{-1})

d = Diameter of the rotation (m)

Form the equation it is obvious that when the speed of rotation increases the C.E increase and when the diameter of the rotation increases the C.E increases. In a larger centrifuge, doubling the speed the centrifugal effect is quadrupling, and a high speed rotation of a small centrifuge gives a very high value of effect.

In this present experiment, we are demonstrating how the centrifugal effect changes with diameter and speed of rotation of the centrifuge.

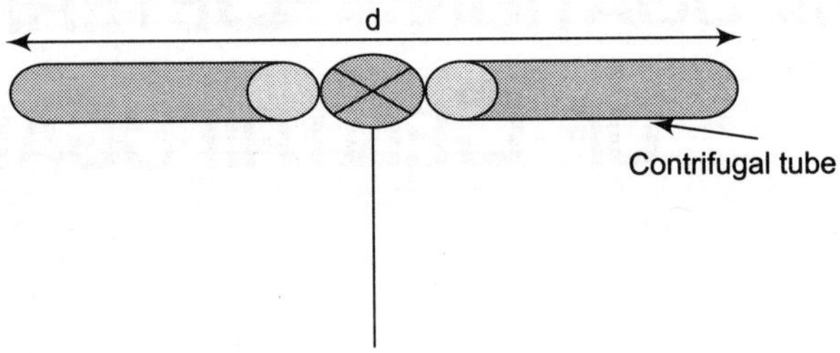

Fig. 36.1: Determination of diameter of rotation.

I. Preparation of emulsion

1. Emulsion is prepared with the help of motor and pestle.
2. The standard formula for the emulsion is

   Castor oil = 37.5 ml

   Acacia = 10.9 gm

   Water to produce = 100 ml.
3. Transfer the prepared emulsion into a clean beaker.
4. Select and thoroughly clean two centrifuges. (big and small)
5. Find out the diameter of rotation for both the centrifuge. (see Fig. 36.1)
6. Fill equal amount of the emulsion in the centrifugal tubes of both the centrifuges. (The volume can be selected according to the capacity of the tube (normally 15–50 ml)
7. Arrange the samples properly in the centrifuge to avoid the accidental breakage during operation.
8. Allow to operate the centrifuges at 500 RPM for 10 min.
9. Stop the centrifuges and wait till the rotation stops.
10. Open the lid and take out the tubes.
11. Find out the amount of the oil separated.
12. Calculate the C.E using the equation.
13. Pour out the content present in the centrifugal tubes and clean it properly.
14. Repeat the steps 6 & 7.
15. Allow to rotate the centrifuge at 1000 rpm for 10 min.
16. Calculate the amount of the oil separated and C.E.
17. The same steps (6, 7, 9 – 12) is repeated for RPM 2000, 3000, 5000.
18. The values are tabulated.
19. Graphs are drawn as shown in the model graphs.

OBSERVATIONS AND CALCULATIONS

Table 36.1: Determination of Centrifugal Effect

Sl No	*Rpm (s^{-1})*	*Diameter of rotation (n) (m)*	*Centrifugal effect (m/s^2)*	*Volume of oil separated (ml)*
Big centrifuge	500 1000 2000 3000 5000			
Small centrifuge	500 1000 2000 3000 5000			

Note: Volume of oil separated is the average volume of oil present in all the tube of the centrifuge after each operation.

RESULT

Performed the Centrifugal separation of emulsion at different speed of rotation.

The centrifugal effect produced during the separation of emulsion in a centrifuge with diameter of rotation d =...... with another centrifuge having d = ... at different speeds of rotation is compared. It is found that as the speed (rpm) increases the C.E .... and as the '*n*' increase the C.E ....The volume of oil separated ....with increase in the RPM .

EXPERIMENT 37

SIMPLE DISTILLATION—ETHANOL WATER SYSTEM

AIM

To perform the simple distillation of ethanol water system.

To verify the Material balancing equation and Rayleigh equation.

REQUIREMENTS

Distillation assemble	(1)
Heating mantle (heat source)	(1)
Thermometer	(1)
Conical flask	(1)
Specific gravity bottle	(1)
Weighing Balance	(1)
Pumice stone	(3–4)
Ethanol	
Water	

PRINCIPLE

Distillation is a method of separating mixtures based on their differences in volatilities, into its component fractions of desired purity, by the application and removal of heat. Distillation is based on the fact that the vapour of a boiling mixture will be richer in the components that have lower boiling points (highly volatile). Therefore, when this vapour is cooled and condensed, the condensate will contain more volatile components. At the same time, the original mixture will contain more of the less volatile material. Simple distillation is usually used to separate liquids whose boiling points differ greatly or to separate liquids from non volatile solids or oils. This experiment demonstrates the separation of ethanol from water. As the distillation proceeds the vapour will be rich in ethanol which will be cooled and collected in a receiver. The process is continued till all the possible amount of the ethanol is distilled off. The Rayleigh Equation is useful in the analysis of simple distillation, as it shows how the concentration and quantity are related. As the simple distillation is unsteady in nature, the equation is based on a differential approach to explain the changes in concentration with time.

$$\boxed{\int_{x_2}^{x_1} dx/(y-x) = \ln S_1/S_2} \quad \text{... (Eq. 37.1)}$$

where,

S_1 = Moles of more volatile component in the feed

S_2 = Moles of more volatile component left in the still

x_1 = Starting mole fraction of more volatile component in the still

x_2 = Final mole fraction of the more volatile component in the still

The integral S_1/S_2 can be determined by graphical method. With the help of $x - y$ data, a plot is drawn x verses $1/y - x$ Draw vertical lines at the limits x_o and x and determine the area enclosed. This is equal to ln (S_1/S_2). (Refer to Fig. 37.3).

Material quantities, as they pass through processing operations, can be described by material balances. Such balances are statements of the conservation of mass. If there is no accumulation, what goes into a process must come out. Material balances are fundamental to the control of processing, particularly for the control of yield of the products. The law of conservation of mass is the basis to what is called as mass or a material balance. The total material balance for the more volatile component during the distillation process can be determined by the following equation,

$$\boxed{F = D + W} \quad \text{... (Eq. 37.2)}$$

where,

F = Number of moles of ethanol in the feed

D = Number of moles of ethanol in the distillate

W = Number of moles of ethanol in the still

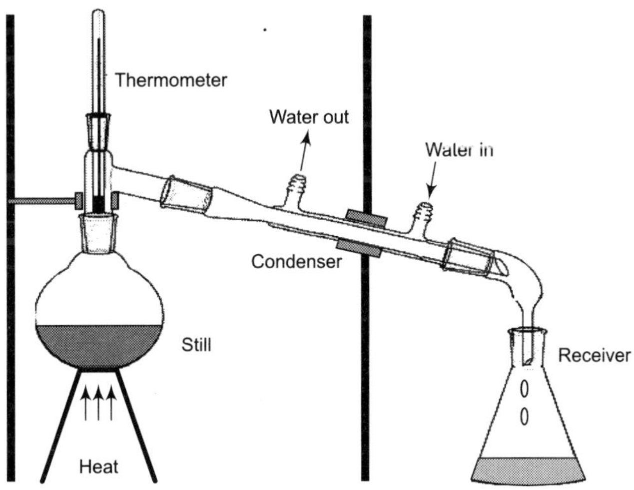

Fig. 37.1: Simple distillation assembly.

PROCEDURE

Preparation of standard curve of Mole fraction Vs Specific gravity

Method I

1. Arrange two burettes in burette stands.
2. Fill one burette with water 100 ml and the other with ethanol 100 ml.
3. Care should be taken to minimize the evaporation of ethanol.
4. Prepare different concentrations of feed (by taking 3, 6, 9, 12, 15, 18, 21, 24, 27, and 30 ml of ethanol and corresponding amount of water to produce the final volume 30 ml.)
5. Find out the specific gravity of each preparation with the help of specific gravity bottle.
6. Find out the mole fraction of ethanol present in all the preparations.
7. Prepare the calibration curve by taking mole fraction of ethanol in X axis and specific gravity in the Y axis.

Method II

Standard table (Alcoholimetric) showing the relationship between % V/V and Density are available in Appendix X.

Simple Distillation

1. Mix 50 ml of ethanol and 50 ml of water in beaker.
2. Find out the specific gravity of the mixture and find out the mole fraction x_F (x_1) from the standard curve or standard table.
3. Find out the number of moles of ethanol present in the feed.
4. Add this mixture (100 ml) into the still (Round bottom flask) of the distillation assemble.
5. Arrange the distillation assemble as shown in Fig. 37.1.
6. Start the distillation and note down the temperature in the thermometer when the mixture starts to boil (a °C).
7. Check the temperature in the thermometer. (It will show a slow change in temperature until all the possible amount of ethanol distill off.)
8. The temperature starts to increase suddenly when the entire possible ethanol is distilled off. At this point, stop the distillation.
9. Wait for 5 min and carefully dismantle the distillation unit.
10. Cool the receiver and still under running water.
11. Find out the specific gravity of the distillate and residue.
12. Find out the mole fraction of distillate as well as residue with the help of standard graph. Mark it as x_D and x_W (x_2) respectively. Mark the x_F (x_1) this can be done with the help of standard table.
13. Find out the relative volatility (a).
14. Plot a graph by taking the standard value of x (mole fraction of more volatile component in the liquid phase) and y (mole fraction of the more volatile component in the vapor phase) (Refer to Appendix XII)
15. Find out $y - x$ and $1/y - x$.

16. Plot another graph by taking x on X-axis and $1/y - x$ on Y axis (see the model graph).
17. Mark the x_1 and x_2 in the graph.
18. Find out the area under the curve from x_1 to x_2.
19. The area under the curve from x_1 to x_2 is equal to ln S_1/S_2.
20. Find out S_2 (*i.e.*, W) and substitute in the material balancing equation to determine the moles of ethanol distilled off (D).
21. Thus the amount of ethanol remaining in the still undistilled, and the amount of ethanol distilled can be determined with the help of Rayleigh equation and material balancing equation.

OBSERVATIONS AND CALCULATIONS

I. Preparation of standard graph (Specific gravity Vs Mole fraction)

Table 37.1: Determination of Mole Fraction of Ethanol

Sl No	*Volume of ethanol (ml)*	*Volume of water (ml)*	*Weight of specific gravity bottle + mixture W2(gm)*	*Weight of the mixture (gm)*	*Specific gravity of the mixture (gm/cm³)*	*Mole fraction of ethanol*
1	3	27				
2	6	24				
3	9	21				
4	12	18				
5	15	15				
6	18	12				
7	21	9				
8	24	6				
9	27	3				
10	30	0				

Determination of density (Refer appendix IV)

Weight of the empty specific gravity bottle $\mathbf{W_1}$ = ... gm

Weight of the bottle + water $\mathbf{W_2}$ = ... gm

Weight of bottle + Mixture $\mathbf{W_3}$ = ... gm

Weight of the mixture W_4 $(\mathbf{W_3 - W_1})$ = ... gm

$$\text{Density of the mixture} = \frac{\text{Weight of the mixture at } t\,°\text{C } (W_3 - W_2)}{\text{Weight of water at } t\,°\text{C } (W_3 - W_2)} \times \text{Density of water at } t\,°\text{C}$$

Molecular weight of ethanol = ...

Molecular weight of water = ...

Specific gravity of the original mixture = ...g/ml

$$\text{Mole fraction of the mixture} = \frac{\dfrac{(\text{Volume of ethanol} \times \text{Specific gravity of the mixture})}{\text{Molecular weight of ethanol}}}{\left[\dfrac{(\text{Volume of ethanol} \times \text{Specific gravity of the mixture})}{\text{Molecular weight of ethanol}}\right] + \left[\dfrac{(\text{Volume of water} \times \text{Specific gravity of the mixture})}{\text{Molecular weight of water}}\right]}$$

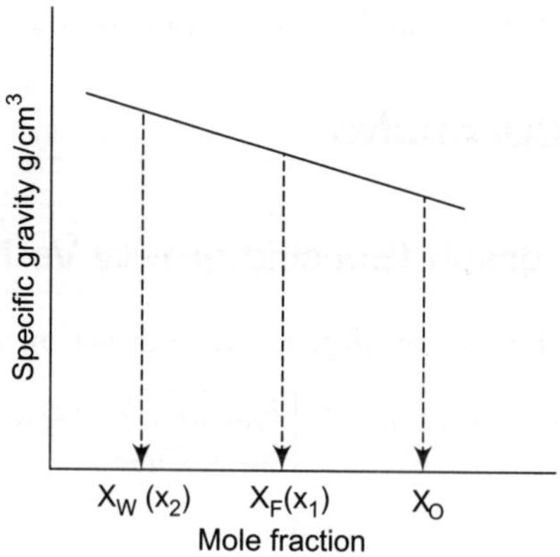

Fig. 37.2: Model graph Mole fraction Vs Specific gravity.

II. Determination of ethanol remaining in the still

Graphical method

Determination of 1/*y* – *x* of ethanol

Table 37.2: Determination of 1/*y* – *x* of Ethanol

Sl No	*x*	*y*	*y* – *x*	1/*y* – *x*

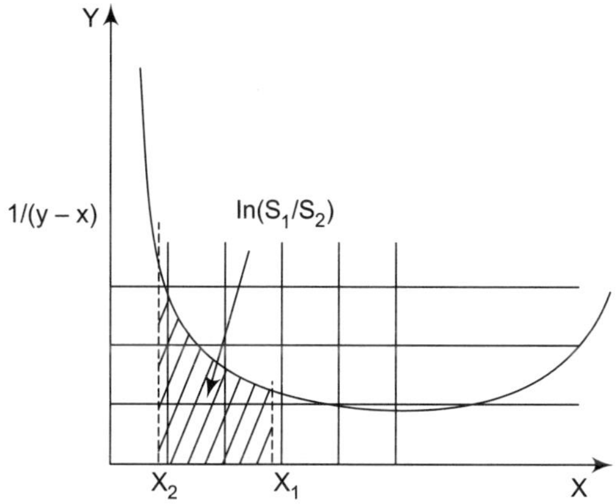

Fig. 37.3: Model graph [x Vs 1/(y – x)].

III. Determination of area under the curve

Area of the triangle = ½ $b \times h$

Area of the rectangle = $l \times h$

Total area under the curve = area of the triangle + area of the rectangle

Or

Area of the trapezoid = ½ b (sum of two parallel sides)

Area under the curve = …………

$(\ln S_1/S_2)$ = …………

$\ln S_1/S_2$ = …a………

$\mathbf{S_1/S_2} = e^a$

$= 2.717828^a$

= …b…………

Density of ethanol = ………..gm/cm^3

$$S_1\text{ (Moles)} = \frac{50 \times \text{Density of ethanol}}{\text{Molecular weight of ethanol}}$$

$\mathbf{S_2}$ (amount of ethanol remaining in the still) = S_1 / b

= …………moles

Number of moles of ethanol distilled **(D)** = F – W (*i.e.*, $S_1 - S_2$)

D= ……………..Moles

REPORT

The simple distillation of ethanol is performed. The amount of ethanol distilled and remaining in the still is determined with the help of Rayleigh and material balancing equation.

EXPERIMENT 38

DETERMINATION OF BOILING POINT OF THE COMPONENTS IN A LIQUID MIXTURE BY SIMPLE DISTILLATION

AIM

To determine the boiling point of components in a binary liquid mixture—ethanol water.

REQUIREMENTS

Distillation assemble	(1)
Heating mantle (heat source)	(1)
Thermometer	(1)
Conical flask	(1)
Pumice stone	(3-4)
Ethanol	
Water	

PRINCIPLE

Distillation is a method of separating mixtures based on differences in their volatilities into its component fractions of desired purity, by the application and removal of heat. The distillation methods are selected depending upon the types of mixture to be separated.

1. If the components in the mixture have greater difference in their individual boiling points, simple distillation can be selected.
2. If the difference in the boiling points is less, fractional distillation is preferred.
3. If it is to separate immiscible component from the mixture having highboiling point, steam distillation is preferred.
4. If a thermolablie substance is to be separated from the mixture vacuum distillation can be used.

This experiment is designed to determine the boiling point (range) of more volatile component during distillation of a liquid mixture of ethanol water system. During the distillation the high volatile component starts to boil first and separated from the still and collected in the receiver. The temperature of the mixture is noted when the mixture starts to boil and the more volatile component starts to evaporate. Throughout the experiment continuous monitoring of the temperature is to be done. When the entire high volatile component is distilled off one can observe the sudden change in the temperature readings in the thermometer. This will be the boiling point of the second component (less volatile)/azeotrope. The temperature readings can be corrected for barometric pressure using the following expression,

$$t_1 = t_2 + \mathrm{K}\,(101.3 - b) \quad \text{... (Eq. 38.1)}$$

where,

t_1 = Corrected temperature (°C)
t_2 = Measured temperature (°C)
b = Barometric pressure in KPa at the time of the determination,
K = Correction factor

Table 38.1: K Values

Boiling point °C	*K*
<100	0.30
100 to 140	0.34
140 to 190	0.38
190 to 240	0.41
>240	0.45

PROCEDURE

1. Accurately measure 50 ml of water and 50 ml of ethanol.
2. Carefully add this into the round bottom flask.
3. Add 1-2 pieces of pumice stone to the flask.
4. Arrange the distillation assemble as shown in the (Fig. 37.1) previous experiment.
5. It is customary to distill from left to right.
6. Start the distillation.
7. Note down the temperature in the thermometer when the first drop of distillate is collected in the receiver.
8. Continue to record the temperature after each successive 2 ml of distillate is collected (or at an interval of specific time *e.g.*, 2 min, 5 min).
9. Note down 5 readings of temperature.
10. Monitor the temperature throughout the distillation and note down the temperature when the sudden change in the temperature is observed.
11. Continue to record the temperature after each successive 2 ml of distillate is collected. (Note down 5 readings of temperature)
12. Report the findings.

Table 38.2: Determination of Boiling Temperature of the Components

Sl No	*Components*	*Time (min)*	*Volume collected (ml)*	*Observed Temperature (°C)*	*Corrected temperature (°C)*	*Average temperature (°C)*
1	High volatile component					
2						
3						
4						
5						
6	Low volatile component/ Azeotrope					
7						
8						
9						
10						

REPORT

The boiling points of components in an ethanol water system is determined by simple distillation method.

EXPERIMENT 39

DETERMINATION OF DISTILLATION RANGE IN A LIQUID MIXTURE OF ETHANOL WATER—SIMPLE DISTILLATION

AIM

To determine the distillation range of components in a binary liquid mixture – ethanol water by simple distillation.

REQUIREMENTS

Distillation assemble	(1)
Heating mantle (heat source)	(1)
Thermometer	(1)
Conical flask	(1)
Pumice stone	(3-4)
Ethanol	
Water	

PRINCIPLE

The distillation range of a liquid is the temperature interval, corrected to a pressure of 101.3 KPa (1 atm), within which the liquid, or a specified fraction of the liquid, distils under the standard conditions. If the phase rule is applied to the two component system in the distillation flask where two phases (liquid and vapour) are present, it can be shown that there exists two degree of freedom.

$$F = 2 - 2 + 2 = 2$$

Since the pressure is kept constant, the temperature will therefore change as the composition varies in order to maintain the same number of phases *i.e.*, the boiling point of the liquid remaining in the flask changes as all the low boiling point component is distilled off from the still. After that the temperature remains constant.

The present experiments is designed to determine the range of temperature at which the distillation of components is happened. It can be done by recording the temperature at which the first drop of distillate falls into the receiver and monitoring the temperature at which the whole or the prescribed fraction of the liquid, measured at 20°C has distilled at an interval of 2 ml. Determination of distillation range gives the correct idea about the temperature conditions at which the distillation is carried out. This data is used to construct the boiling point diagram.

PROCEDURE

1. Accurately measure 50 ml of water and 50 ml of ethanol.
2. Carefully add this in to the round bottom flask.
3. Add 1-2 pieces of pumice stone to the flask.
4. Arrange the distillation assemble as shown in Fig. 37.1.
5. It is customary to distill from left to right.
6. Start the distillation.
7. Note down the temperature in the thermometer when the first drop of distillate is collected in the receiver.
8. Continue to record the temperature after each successive 2 ml of distillate is collected.
9. Transfer the first 10 ml of distillate into a stoppered glass vial and mark it as F_1. (Standard flask of 20 ml can also be used for this purpose).
10. Continuously note down the temperature and collect the next 20 ml of distillate.
11. Transfer this to another glass vial and mark it as F_2.
12. Similarly the 3rd fraction and 4th fraction of distillate is also collected and mark it as F_3 and F_4 (20 ml each)
13. Stop the distillation after collecting total 80 ml of distillate.
14. Report the distillation range of the ethanol-water mixture.

OBSERVATIONS AND CALCULATIONS

Table 39.1: Determination of Temperature During Distillation

Volume collected (ml)	*Temperature (°C)*	*Volume collected (ml)*	*Temperature (°C)*	*Volume collected (ml)*	*Temperature (°C)*	*Volume collected (ml)*	*Temperature (°C)*
2		22		42		62	
4		24		44		64	
6		26		46		66	
8		28		48		68	
10		30		50		70	
12		32		52		72	
14		34		54		74	
16		36		56		76	
18		38		58		78	
20		40		60		80	

Table 39.2: Determination of Temperature Range

Sl No	*Volume collected*	*Temperature range*
1	10	
2	20	
3	20	
4	20	

REPORT

The distillation range of the given mixture of ethanol-water is ....

EXPERIMENT 40

STEAM DISTILLATION—DETERMINATION OF EFFICIENCY OF DISTILLATION

AIM

To perform the steam distillation of clove oil.

To determine the percentage recovery of the clove oil during steam distillation.

To determine the % efficiency of steam distillation.

REQUIREMENTS

Steam distillation assemble	(1)
(Round bottom flask, Measuring cylinder, Steam can, Adaptors, Bend tubes, Burette stands, Thermometer)	
Weighing balance	(1)
Beakers (250 ml)	(2)
Separating funnel	(1)
Clove oil	
Water	

PRINCIPLE

Steam distillation refers to a process in which live steam is in direct contact with the distilling system in either batch or continuous operation. The basis of steam distillation rest on the fact that water forms immiscible mixtures with most organic substances and these mixtures will boil at a temperature below that of either water or the other materials. This is a very desirable feature, especially if the organic compound has a high boiling point at which it may be unstable or decompose. As long as liquid water is present, the high boiling component will vaporize at a temperature well below its normal boiling point. Steam distillation is mainly used for:

1. Separating relatively small amounts of volatile impurity from a large amount of material.
2. Separating appreciable quantities of higher-boiling materials.
3. Recovering high-boiling materials from small amounts of impurity which have a higher boiling point.
4. Distilling thermally unstable or components react with other components associated with it at the boiling temperature.
5. The materials cannot be distilled by indirect heating even under low pressure because of the high boiling temperature.
6. Distillation where direct-fired heaters cannot be used because of fire hazards.

In this present experiment the separation of clove oil from water is done by steam distillation. The clove oil (eugenol) is having a molecular weight of 164.20 and boiling point of 254 °C. When this is mixed with water the boiling point is reduced and the mixture starts boiling at fairly low temperature. Steam is widely used because of its energy level, cheapness, and availability. The weight of clove oil which distills along with water can be determined with the help of the equation,

$$\boxed{W_2 = W_1 (M_2P_2)/(M_1P_1)} \quad \text{... (Eq. 40.1)}$$

where,

W_2 = Weight of clove oil (g)
W_1 = Weight of water (g)
M_1 = Molecular mass of water
M_2 = Molecular mass of clove oil
P_1 = Partial pressure of water (mmHg)
P_2 = Partial pressure of clove oil (mmHg)

The partial pressure of the clove oil at the boiling point can be determined by,

$$P_{total} = P_{water} + P_{clove\ oil} \quad \text{... (Eq. 40.2)}$$

$$\boxed{\text{The efficiency of the distillation = Amount recovered/Amount used} \times 100} \quad \text{... (Eq. 40.3)}$$

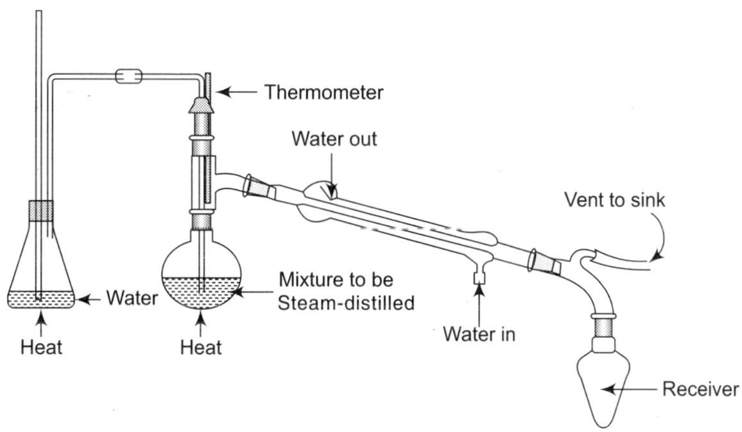

Fig. 40.1: Steam distillation assemble.

PROCEDURE

1. Measure 15 ml of clove oil, and 50 ml of water.
2. Add both the liquids in the round bottomed flask.

3. Arrange the steam distillation assemble as shown in Fig. 40.1.
4. Fill the steam can with water and heat it so that a continuous flow of steam is generated.
5. As the steam entered into the still the mixture is heated.
6. Continuously monitor the temperature in the thermometer.
7. The temperature is noted when the mixture starts to boil.
8. The mixture starts boil and distills off to the receiver via condenser.
9. Note down the temperature as well as the amount collected at an interval of 5 min.
10. Distillation is carried out till < 5 ml of mixture is remaining in the RB flask.
11. It is advised not to distill the content in the flask to dryness.
12. Separate the contents in the receiver with the help of a separating funnel.
13. Find out the weight of the clove oil as well as water individually.
14. Find out the practical recovery of clove oil.
15. Determine the efficiency of recovery.
16. Find out the rate of distillation.

OBSERVATIONS AND CALCULATIONS

Average rate of distillation = ...

Boiling point of the mixture = ...*a*......°C

Partial pressure of water at a °C = ...*b*....

(Refer to Appendix XI)

Partial pressure of eugenol $= P_{total} - P_{water}$

= 760 – ....

= ......*c*......

Determination of Rate of Distillation

Table 40.1: Determination of Amount of Distillate Collected with Time

Sl No	*Time (min)*	*Temperature (°C)*	*Amount collected (ml)*	*Rate of distillation (ml/min)*

Determination of Theoretical Recovery

Weight of clove oil (Initial) = 15 × Weight per ml of clove oil

= ……*x*…gm……

Weight of clove oil distilling with 1 gm of water $W_2 = W_1 (M_2P_2)/(M_1P_1)$

$\mathbf{W_2} = 1 \times (164.20 \times c)(18 \times b)$

= …………gm

Weight of water Separated (distilled) = ………*d*……..gm

Weight of clove oil distilled ($W_2 \times d$) = ……..*y*………gm

Theoretical recovery = $y/x * 100$

= ……………%

Determination of Practical Recovery

Weight of clove oil separated (Distilled) = ……*e*……….gm

Weight of water separated = ……………gm

Practical recovery of clove oil = $e/x * 100$

The efficiency of the distillation = Practical recovery/Theoretical recovery × 100.

= …………………..

REPORT

The steam distillation is performed.

The weight of clove oil distilled with I gm of water = …………

The efficiency of distillation = ………. %

The average rate of distillation = ……….ml/min

EXPERIMENT 41

VACUUM DISTILLATION—EFFECT OF VACUUM ON DISTILLATION OF ASPIRIN REACTION MIXTURE

AIM

To perform the vacuum distillation at varying pressure.

To find out the rate of distillation at varying pressure (vacuum).

To determine the effect of vacuum on the distillation of thermo labile drug (aspirin).

REQUIREMENTS

Vacuum distillation assemble (Vacuum pump, Claisen flask, Round bottom flask, Measuring cylinder, Separating funnel, Stop cock, Adaptors, Bend tubes Thermometer, Condenser)

Heating mantle	(1)
Tripod stand	(2)
Burette stands	(3)
Ethanol	
Water	
Aspirin	

PRINCIPLE

Vacuum distillation is a distillation process used for materials that have very high boiling-points or are likely to decompose at their boiling-points. Vacuum distillation is a distillation at reduced pressure. Since the boiling point of a compound is lower at a lower external pressure, the compound will not have to be heated to as high a temperature in order for it to boil. The vacuum is provided either by a water aspirator or by a mechanical pump, and the vacuum so formed causes the materials being heated to boil at a temperature lower than they would under normal atmospheric pressure. This enables distillation to take place at lower temperatures. A reliable vacuum can be obtained using a motor-driven oil pump. Vacuum distillation is used with or without heating the solution. Applications of Vacuum distillation are,

1. When liquids to be distilled have high atmospheric boiling points or chemically change at temperatures near their atmospheric boiling points.
2. Temperature sensitive materials (such as beta carotene) also require vacuum distillation to remove solvents from the mixture without damaging the product.
3. Compared to steam distillation there is a lower level of residue build up.

Aspirin is a drug which will undergo hydrolysis to form salicylic acid. The degradation increase with increase in temperature. During the preparation of aspirin, the formed aspirin should be recovered from the reaction mixture by distillation. As the drug is temperature sensitive the recovery process cannot be done with the help of heat. As vacuum distillation works under a low temperature, it can decrease the boiling point of the reaction mixture. So during distillation the drug can be protected from degradation. By increasing the vacuum, one can observe the increase in the rate of distillation and lesser degradation of the drug.

During the distillation the total vapour pressure for the system can be determined by:

$$\boxed{P_{total} = pA + pB} \quad \text{... (Eq. 41.1)}$$

where,

pA = Partial pressure of the component A

pB = Partial pressure of component B

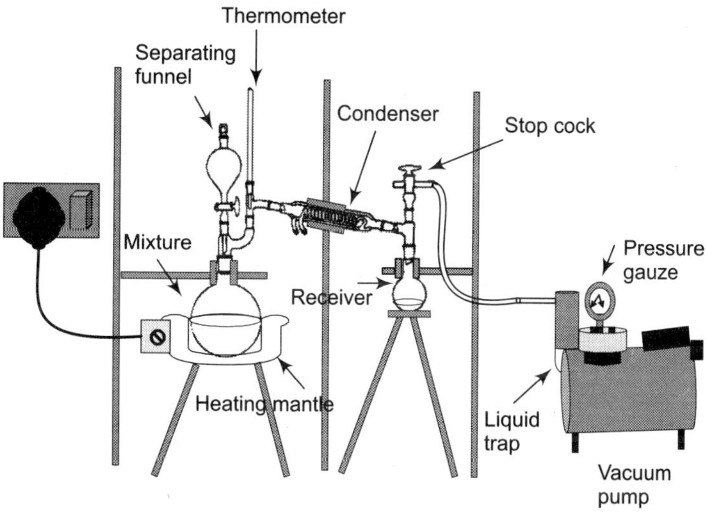

Fig. 41.1: Arrangement for vacuum distillation of aspirin.

PROCEDURE

1. Accurately measure 30 ml of ethanol and 70 ml of water.
2. Weigh 100 mg of aspirin.
3. Dissolve it in the 30 ml of ethanol and mix with 70 ml of water.
4. 5 ml of the mixture is preserved for the analysis for the original concentration of aspirin present.

5. Add the rest of the mixture into the claisen flask and arrange the distillation apparatus as shown in Fig. 41.1.
6. Open the stop cock of the receiver and start the vacuum pump.
7. The decrease in the pressure can be noted in the pressure gauge.
8. After attaining a vacuum of 600 mm Hg the stop cock is closed and the vacuum pump is stopped so that the vacuum is maintained.
9. Heat the mixture with the help of heating mantle/water bath/oil bath.
10. When the total pressure of the mixture reaches the pressure created (600 mm Hg) boiling stats.
11. Note down the temperature at which the boiling of the mixture occurred.
12. From this point at an interval of 2 min, the boiling point as well as the volume of the distillate collected is noted. (A calibrated measuring cylinder can be used as the receiver. It is not advisable to use the normal measuring cylinder as the adapter does not fit tightly to the normal one.)
13. Find out the partial pressures of the individual components using the equation 41.1. (Refer to Appendix XI)
14. The distillation is suspended when the temperature of the mixture suddenly rises because of distillation of the ethanol water azeotrope.
15. Wait for five minutes.
16. Draw the standard graph for the salicylic acid (Refer to Appendix IX)
17. Dismantle the apparatus and take out 1 ml of sample from distillate and find out the amount of salicylic acid formed by UV spectrophotometer.
18. Take out 1 ml of residue after proper dilution and find out the amount of salicylic acid formed in the residue (Refer to Appendix IX)
19. Find out the total salicylic acid formed by combining the amount of salicylic acid present in the distillate and in the residue.
20. Find out the total aspirin decomposed and undecompensed.
21. Find out the percentage of aspirin protected after distilling it at a pressure of 600 mm Hg.
22. Determine the rate of distillation.
23. The same procedure is repeated for vacuum of 400 and 800 mm Hg.
24. Report the findings.
25. Find out the effect of vacuum on rate of distillation.

Precaution: Two problems are usually encountered in a vacuum distillation. The volume of vapour formed from a given amount of liquid is pressure dependent; *i.e.*, the volume of vapour formed from one drop of liquid will be about 30 times as great at 25 torr as it was at 760 torr. As a result, serious bumping may occur. Boiling chips generally do not help much at the reduced pressures. Some of the bumping can be overcome with the use of a magnetic stir bar. The second problem is also related to the larger volume of vapour at lower pressure. The velocity of the vapour entering the column is greatly increased. This creates a greater pressure in the column than is registered on the manometer. Maintaining a slow, steady rate of distillation can minimize this difference in pressure.

OBSERVATIONS AND CALCULATIONS

Amount of salicylic acid present
in the mixture before distillation = (C) before distillation × 100
= ...*x*.......mg

Amount of decomposed aspirin
present in mixture before distillation = $x \times 1.3$
= ......*y*.....mg

Amount of original concentration = $100 - y$
of aspirin = ...Z..... mg

Table 41.1: Determination of Concentration of Salicylic Acid

	Pressure at which distillation is carried out (mm Hg)	Volume taken for analysis(ml) A	Volume of ferric chloride solution (4%w/v)	Volume of distilled water	Dilution factor	Concentration of salycylic acid in the solution mg/ml (B)	Absorbance at 547 nm I_s	Concentration of SA in the sample taken (B X dilution factor in mg/ml) (C)
Distillate	400		5 ml					
	600		5 ml					
	800		5 ml					
Residue	400		5 ml					
	600		5 ml					
	800		5 ml					
Before distillation			5 ml					

Table 41.2: Determination of Amount of Aspirin Protected

	Pressure at which distillation is carried out (mm Hg)	Concentration of SA in the sample taken in mg/ml (C)	Total Volume collected (ml) (D)	Con of SA in volume collected C X D = E mg	amount of aspirin decomposed E X 1.3 = F mg	Amount of aspirin un decomposed Z – F =H mg
Distillate	400					
	600					
	800					
Residue	400					
	600					
	800					

Table 41.3: Comparison of % Aspirin Protected with Increase in Vacuum

Sl No	*Vacuum pressure (mmHg)*	**Concentration of aspirin undecomposed in distillate (I) (mg)*	***Concentration of aspirin undecomposed in Residue (II) mg*	*Total amount of aspirin undecomposed After distillation (III) = (I) + (II) mg*	*% of aspirin protected (III)/Z x 100*

* H from the table 41.2

** H from the table 41.2 for residue at corresponding pressure

Table 41.4: Determination of Rate of Distillation

Sl No	*Pressure (mm Hg)*	*Time (min)*	*Volume of distillate collected (ml)*	*Rate of distillation (ml/min)*

Table 41.5: Determination of Partial Pressure of Individual Components

Sl No	*Produced Vacuum pressure (mm Hg)*	*Boiling point of the mixture (°C)*	*Partial pressure of ethanol at the boiling point of the mix mmHg (p_A)*	*Partial pressure of water at the boiling point of the mixture (mmHg) (p_B)*	*Total vapor pressure of the mixture mmHg (P_{total})*

REPORT

Vacuum distillation is performed and the effect of vacuum on the distillation of thermo labile drug is demonstrated.

With increase in vacuum, the percentage of aspirin protected ....

With increase in vacuum, the rate of distillation ....

With increase in vacuum the boiling point of the mixture ....

EXPERIMENT 42

DETERMINATION OF NUMBER OF THEORETICAL PLATES—FRACTIONAL DISTILLATION

AIM

To determine the number of theoretical plates while performing the fractional distillation.

To determine the HETP (Height equivalent to one theoretical plate).

REQUIREMENTS

Fractional distillation assemble (1)

(Fractionating Column, Round bottom flask, Condenser, adaptor, receiver, thermometer)

Pipes (2)

Specific gravity bottle (2)

Heating mantle/ burner (1)

Tripod stands (2)

Burette stands (2)

Conical flask (1)

Water

Ethanol

PRINCIPLE

Fractional distillation is a process by which components in a chemical mixture are separated according to their difference in the boiling points. Vapors from a boiling solution are passed along a fractionating column. The temperature of the column gradually decreases along its length. Components with a higher boiling point condense on the column and return to the solution; components with a lower boiling point pass through the column and are collected as fractions.

Fractionating columns are of two types, plate columns and packed columns. In this experiment simple spiral packed column is used.

Theoretical plates can be defined as a plate holding a pool of liquid in which the average composition of the liquid vaporizes is in equilibrium with the average composition of the liquid condensing. Theoretical plates can be calculated with the help of Fenske equation:

$$\mathbf{N + 1 = log \frac{(x/1 - x)\ distillate\ (1 - x/x)\ residue}{Log\ \alpha_{av}}} \qquad \text{... (Eq. 42.1)}$$

where,

N = Total number of theoretical plates

α = Relative volatility of more volatile component with respect to less volatile component

x = Mole fraction of more volatile compound

$\alpha_{av} = p_A / p_B$

p_A = Vapour pressure in mm Hg of more volatile component

p_B = Vapour pressure in mm Hg of less volatile component

Determination of theoretical plates is helpful to find out the feeding position and the reflux during the fractional distillation. This will also give an idea about the plate efficiency. The HETP can be determined with the help of the following equation:

$$\mathbf{HETP = H/N} \qquad \text{... (Eq. 42.2)}$$

where,

H = Height of the packing (inch)

N = Number of theoretical plates

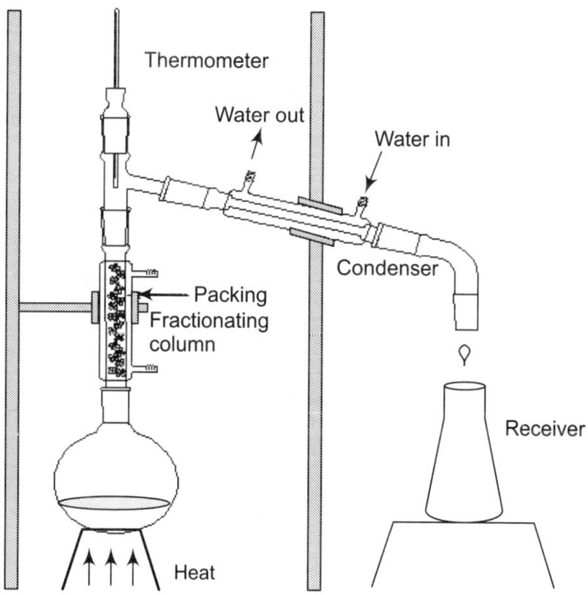

Fig. 42.1: Fractional distillation assemble.

PROCEDURE

1. Exactly measure 100 ml of water and 100 ml of ethanol.
2. Add both the liquids in a round bottom flask and put 2-3 pumice stone to minimize the bumping.
3. Find out the specific gravity of the mixture.
4. Attach the fractionating column to the round bottom flask and arrange the distillation assemble. (see Fig. 42.1)
5. Heat the mixture with the help of heating mantle or any other heating sources.
6. Note down the temperature when mixture starts to boil.
7. The more volatile component starts to collect at the receiver.
8. Wait till the 50 ml of distillate is collected in the receiver.
9. Stop heating by removing the heating source.
10. Slowly dismantle the assemble.
11. Find out the specific gravity of the distillate as well as liquid remaining in the still.
12. Note down the boiling point of the distillate (pure ethanol).
13. Find out the mole fraction of the distillate and bottom product.
14. Find out the α_{av} of the mixture at their boiling temperature (find out the vapour pressure values from the table given in appendix XI).
15. Substitute in the equation and report the number of theoretical plates.

OBSERVATIONS AND CALCULATIONS

Table 42.1: Determination of Mole Fraction of Ethanol & Water

Sl No	*Sample*	*Specific gravity (g/ml)*	*Moles of water*	*Moles of Ethanol*	*Mole fraction of water*	*Mole fraction of ethanol (x)*
1	Initial mixture					
2	Distillate					
3	Bottom product					

Specific gravity of the original mixture = ...g/ml

Boiling temperature of the initial mixture **(*a*)** = ...°C

Boiling point of the distillate **(*b*)** = ...°C

Vapour pressure of ethanol at **(*a*)** °C = ...

(Refer the appendix XI) = ...

Vapour pressure of ethanol at **(*b*)** °C = ...

Average vapour pressure of ethanol p_A = ...

Vapour pressure of water at (*a*) °C = ...

Vapour pressure of water at (*b*) °C = ...

Average vapour pressure of water p_B = ...

$\alpha_{av}(p_A/p_B)$ = ...

Total height of the packing = ...inch

Determination of Number of Moles and Mole Fractions

Number of moles = Gram of solute/Molecular weight

Number of moles of ethanol present in the initial mixture

$$= \text{Gram of ethanol/ MW}$$

$$= \frac{100 \times \text{Density of ethanol}}{\text{Molecular weight of ethanol}}$$

$$= \ldots\ldots\ldots\ldots\ldots\ldots$$

Number of moles of water present in the initial mixture

$$= \text{Gram of water/MW}$$

$$= \frac{100 \times \text{Density of water}}{\text{Molecular weight of water}}$$

$$= \ldots\ldots\ldots\ldots\ldots\ldots$$

$$\text{Mole fraction of ethanol present in the initial mixture} = \frac{\text{Moles of ethanol}}{\text{Moles of ethanol + Moles of water}}$$

$$= \ldots\ldots\ldots\ldots\ldots\ldots\ldots$$

$$\text{Mole fraction of water present in the initial mixture} = \frac{\text{Moles of water}}{\text{Moles of water + Moles of Ethanol}}$$

$$= \ldots\ldots\ldots\ldots\ldots\ldots\ldots\ldots\ldots$$

Number of moles of ethanol present in the distillate = Gram of solute / molecular Weight

$$= \frac{50 \times \text{density of ethanol}}{\text{Molecular weight of ethanol}}$$

$$N = \left(\frac{\left[\log\left(x/1-x\right) \text{distillate} \left(1-x/x\right) \text{residue}\right]}{\text{Log } \alpha_{av}} \right) - 1$$

$$= \ldots\ldots\ldots\ldots\ldots\ldots$$

$$\text{HETP} = \text{H/N}$$

$$= \ldots\ldots\ldots\ldots$$

REPORT

The number of theoretical plates present during the fractional distillation of water and ethanol is found to be ....

The HETP of the packed column is ....

EXPERIMENT 43

CRYSTALLIZATION BY COOLING

AIM

To perform the crystallization by cooling.

To compare the effect of type of cooling on yield of the crystals.

REQUIREMENTS

Test tubes	(10)	KNO_3
Test tube stand	(1)	Ice
Hot water bath	(2)	
Thermometers	(2)	
Beaker (250 ml)	(1)	
Stirrer	(1)	
Tripod stand	(1)	
Funnels	(3)	
Filter papers		
Weighing balance		

PRINCIPLE

Crystallization is a process of spontaneous arrangement of particle in to repeatedly arrange geometrical pattern from a supersaturated solution. Crystallization can occurred if solution is in a supersaturated state. The super saturation can be produced by either of these following ways.

1. Super saturation by cooling
2. Super saturation by the evaporation of the solvent
3. Super saturation by adiabatic evaporative cooling

4. Super saturation by adding a substance that reduces the solubility of the substance in the solution.

This experiment is based on crystallization by cooling and is only applicable to the solute whose solubility increases with increase in temperature. Cooling can be done by two methods.

1. Shock cooling
2. Slow cooling

The rate of crystallization during shock cooling is fast and the size of the crystals produced is small. But during crystallization by slow cooling the rate of crystallization is slow, and it will achieve comparatively big crystals. Saturated solutions are prepared by elevating the temperature of the solution. At elevated temperature the solubility exceeds the normal solubility. This solution can be subjected to sudden cooling and slow cooling. During cooling the solubility of the solute decreases and as a result the dissolved solute starts to crystallize. The yield is dependent on the time of contact and the temperature. The yield is calculated in terms of percentage. A graph can also be plotted by taking time on X axis and yield in percentage on Y axis.

PROCEDURE

Crystallization

1. Weigh 150 gm of KNO_3.
2. Take 100 ml of water in beaker.
3. Keep the beaker in a water bath at 60 °C.
4. Slowly Add the KNO_3 in small portions in to the beaker containing water.
5. Stir the solution continuously and Stop the addition of solute when a small portion of the solute remains undissolved at the bottom.
6. Weigh the remaining quantity of the KNO_3 and find out the amount of KNO_3 used to produce the saturated solution.
7. Note down the amount of KNO_3 used to produce a saturated solution at elevated temperature (*i.e., x* gm in 100 ml)
8. Keep the solution in the same water bath for five min to ensure the saturation.
9. Mark & Arrange 10 test tubes in a test tube rack.
10. Add 10 ml of hot saturated solution in to all of the test tubes.
11. Keep the entire test tubes immediately into an ice bath.
12. Take out the first test tube after the commencement of 10 min.
13. Filter the solution present in the test tube (using pre weighed filter paper).
14. The crystals in the filter paper are kept for air drying (a hair dryer can also be used for drying the crystals).
15. After drying the crystals find out the yield. (weight of the filter paper along with crystals-weight of the empty filter paper)
16. Find out the percentage yield. (Amount of crystals produced from 10 ml/amount of solute present in 10 ml) × 100
17. Take out 2nd, 3rd, 4th ....10th test tubes at an interval of 15, 20, 25, 30, ...60 min respectively.
18. The same procedure 13-16 is repeated for all the solutions.

19. A graph is plotted by taking percentage yield at different time interval on Y axis and time in minutes on X axis.
20. Repeat the steps 1-10
21. Now keep the test tubes in the normal atmosphere temperature.
22. After 10 min filter the content present in the first test tube.
23. Repeat the steps 14-16
24. Take out 2nd, 3rd, 4th ....10th test tubes at an interval of 15, 20, 25, 30,...60 min respectively
25. Repeat the steps 13-16.
26. A graph is plotted by taking percentage yield at different time interval on Y axis and time in min on X axis.
27. Compare the two graphs.

OBSERVATIONS AND CALCULATIONS

Shock cooling

Concentration of saturated solution at elevated temperature = ...gm /100 ml

Concentration of solute present in 10 ml of saturated solution at elevated temperature = ...gm

Percentage weight of Crystals produced from

$$10 \text{ ml} = \frac{\text{Weight of crystal Produced at 10 min} \times 100}{\text{Weight of solute present in 10 ml}}$$

(Concentration of solute present in 10 ml at elevated temperature) = ...%

Table 43.1: Determination of Percentage Yield of Crystals (Shock Cooling)

Sl No	*Time (min)*	*Weight of empty filter paper (W_1)*	*Weight of filter paper with crystals (W_2)*	*Weight of crystals $W_3(W_2 - W_1)$*	*% Weight of crystals*
1					
2					
3					
4					
5					
6					
7					
8					
9					
10					

Table 43.2: Determination of Percentage Yield of Crystals (Slow Cooling)

Sl No	*Time (min)*	*Weight of empty filter paper (W_1)*	*Weight of filter paper with crystals (W_2)*	*Weight of crystals W_3(W_2-W_1)*	*% Weight of crystals*
1					
2					
3					
4					
5					
6					
7					
8					
9					
10					

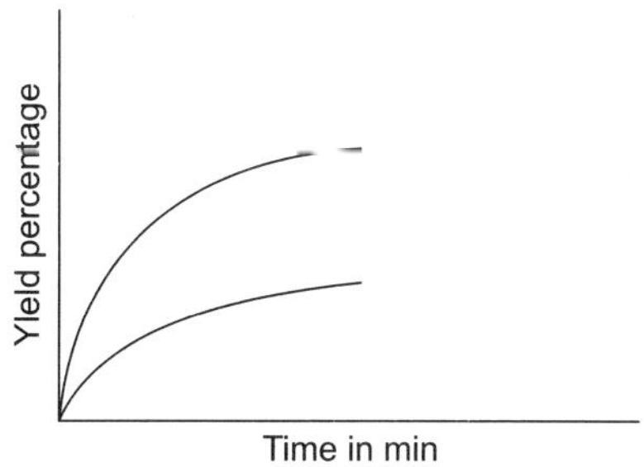

Fig. 43.1: Model graph showing Comparisons of yield obtained by shock cooling and slow cooling with time.

REPORT

The percentage yield of KNO_3 crystals by shock cooling after 45 min =.........

The percentage yield of KNO_3 crystals by slow cooling after 45 min =.........

EXPERIMENT 44

CRYSTALLIZATION—VERIFICATION OF MATERIAL BALANCING EQUATION

AIM

To verify the material balancing equation of crystallization.

REQUIREMENTS

Beaker	(1)
Thermometer	(1)
Stirrer	(1)
Hot air oven	(1)
Water bath	(1)
Tripod stand	(2)
Weighing balance	(1)
Filter paper	
Ice	
Potassium nitrate	

PRINCIPLE

Crystallization is a process of spontaneous arrangement of particle into repeatedly arranged geometrical pattern from a supersaturated solution. Crystallization can occur if solution is in a supersaturated state. The super saturation can be produced by either of these following ways.

1. Super saturation by cooling
2. Super saturation by the evaporation of the solvent
3. Super saturation by adiabatic evaporative cooling
4. Super saturation by adding a substance that reduces the solubility of the substance in the solution.

This experiment is based on crystallization by cooling. This method is only applicable to the solute whose solubility increases with increase in temperature.

A mass balance (also called a material balance) is an application of conservation of mass to the analysis of physical systems. By accounting for material entering and leaving a system, mass flows can be identified which might have been unknown, or difficult to measure without this technique. Under steady state conditions the overall material balancing of any system can be determined from the following equation,

$$\boxed{F = L + C} \quad \text{... (Eq. 44.1)}$$

Or

$$\boxed{F\,X_F = L\,X_L + C\,X_C} \quad \text{... (Eq. 44.2)}$$

where,

F = Feed (liter/hr)
L = Filtrate (liter/hr)
C = Crystals (kg/hr)
X_F = Weight fraction of solute in the feed
X_L = Weight fraction of solute in filtrate
X_C = Weight fraction of solute (crystals) in solid state

PROCEDURE

1. Take 100 ml of water and heat up to 60 °C in water bath.
2. Weigh 200 mg of potassium nitrate (KNO_3).
3. Add KNO_3 to the hot water in small quantities.
4. Stir the solution properly after each addition.
5. Addition of the KNO_3 is continued till a saturated solution is formed. (Some amount of the solute will remain un dissolved at the bottom of the beaker is the indication of attaining the saturation)
6. Weigh the remaining quantity of KNO_3 and note down the quantity of KNO_3 present in the saturated solution.
7. Take the beaker out from the water bath.
8. Immediately place the beaker in ice box for 10 min.
9. The solute starts to crystallize as the temperature decreases.
10. After 10 min filter the crystals produced.
11. Dry the crystals produced under a fan (Hair dryer can also be used).
12. Weigh the crystals when they are completely dried. (C)
13. Measure the filtrate collected.
14. Concentrate the filtrate on a water bath.
15. When it reaches $\frac{1}{4}^{th}$ quantity transfer it into a pre weighed China dish.
16. Heat the China dish till solid residue is obtained.
17. Place the China dish into a hot air oven.
18. Take out the China dish and weigh it.

19. Find out the weight of the residue present.
20. Determine the weight fractions x_F, x_L, x_C.

OBSERVATIONS AND CALCULATIONS

Weight of KNO_3 to produce a saturated solution = 200 – Quantity of KNO_3 remaining after preparing saturated solution

= .... gm

= .... kg

Weight of the crystals = ... gm

= .... kg

Volume of filtrate = ... ml

= .... liter

Weight of the empty China dish (**W_1**) = ... gm

Weight of the China dish + residue (**W_2**) = ... gm

Weight of the residue (**W_3**) = $W_2 - W_1$

= .... gm

= .... kg

Material balancing equation

$$F x_F = L x_L + C x_C$$

F = 0.1 litre

L = Quantity of the filtrate in liter

= ....Liter

C = Weight of the crystals produced (Kg)

= ....Kg

x_F = Amount of the solute present in the feed/Amount of the solute in the feed+ amount of solvent

= ....

x_L = Amount of the solute present in the filtrate/ Amount of filtrate + Amount of solute present

= ....

x_C = Amount of crystals produced/Amount of crystals produced + Amount of crystals produced

= ....

$$F x_F = L x_L + C x_C$$

.... = ... + ...

REPORT

Verified the material balancing equation of crystallization.

EXPERIMENT 45

EVAPORATION—EVALUATION OF FACTORS AFFECTING RATE OF EVAPORATION

AIM

To evaluate the factors affecting the rate of evaporation.

REQUIREMENTS

Heating mantle	(6)
Analytical balance	(1)
Electronic balance	(1)
Beaker 50 ml	(6)
Beaker 150 ml	(6)
Beaker 250 ml	(1)
Beaker 500 ml	(1)
Stirrer	(6)
Measuring cylinder	(1)
Sugar	
Water	

PRINCIPLE

Evaporation is defined as the removal of liquid from a solution by boiling the liquid in a suitable vessel and withdrawing the vapour leaving a concentrate liquid residue. There are so many factors which affect the evaporation. Under this title we are evaluating the effect of temperature, surface area for heat transfer, concentration of the feed and the viscosity of the feed on rate of evaporation. As the temperature provided for evaporation increases, the rate of evaporation is also increases. In this experiment water is allowed to evaporate at different temperatures and the effect of temperature on evaporation rate is determined on account of amount of water lost during the process.

Concentration of the feed which is subjected to evaporation is also affecting the evaporation rate. As the concentration of the feed increases the density as well as the viscosity increases. The boiling point of the solution may also rise considerably so that the boiling temperature of a concentrated solution is considered much higher than that of water (solvent) at the same pressure. Effect of concentration on the rate of evaporation is demonstrated by subjecting solution containing different concentration of any salt in same quantity of water to evaporation.

An Increase in the surface area will increases the heat transferred. So that there will be observed an increase in the rate of evaporation. Number of beakers of different size containing the same quantity of water subjected to evaporation can demonstrate the effect of surface area on rate of evaporation.

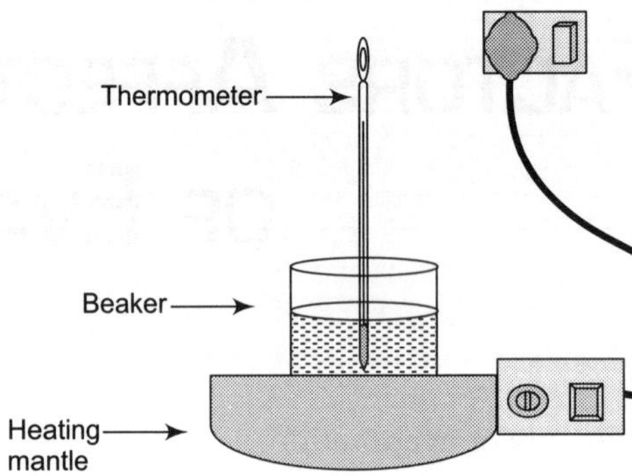

Fig. 45.1: Arrangement for Evaluation of factors affecting evaporation.

PROCEDURE

Effect of Temperature on Evaporation

1. Take 6 beakers of same volume (100 ml).
2. Clean thoroughly and weigh all the beakers individually and record the values as W_1.
3. Mark the beakers I-VI.
4. Arrange 6 different heating mantles and set at different temperatures. (60, 70, 80, 100, 120 and 140 °C)
5. Wait till the mantles acquire the desired temperature.
6. Place 50 ml of water in all the beakers.
7. Weigh all the beakers individually and record the weight as W_2 (Beaker + Water)
8. Place the beakers on the mantles.
9. Keep the beakers on the mantles for at least 30 min.
10. After completion of 30 min carefully take out all the beakers.
11. Allow to cool the water in the beaker.
12. Weigh all the beakers individually and record the weight as W_3.
13. Find out the weight of water evaporated.
14. Find out the rate of evaporation (g/ml).
15. Transfer the water remaining in the beaker to a measuring cylinder and note down the quantity remaining after evaporation (C ml).
16. Calculate the rate of evaporation (ml/min).
17. A graph can be plotted by taking temperature on X axis and rate of evaporation on Y axis.

Effect of Surface Area on Rate of Evaporation

1. Take 4 beakers of different size (50 ml, 100 ml, 250 ml and 500 ml).
2. Clean thoroughly and weigh all the beakers individually and record the values as W_1.
3. Arrange 4 different heating mantles and set the temperature 80 °C.
4. Wait till the mantles acquire the desired temperatures.
5. Place 50 ml of water in all the beakers.
6. Weigh all the beakers individually and record the weight as W_2 (Beaker + Water)
7. Place the beakers on the heating mantles.
8. Heat it for at least 30 min.
9. After completion of 30 min carefully take out all the beakers.
10. Allow to cool the water in the beaker.
11. Weigh all the beakers individually and record the weight as W_3.
12. Find out the weight of water evaporated.
13. Transfer the water remaining in the beaker to a measuring cylinder and note down the quantity remaining after evaporation (C ml).
14. Calculate the rate of evaporation.
15. A graph can be plotted by taking surface area of the beakers in m^2 on X axis and rate of evaporation (ml/min) or (gm/min) on Y axis.
16. Compare the effect of surface area on rate of evaporation.

Effect of Added Substances on Rate of Evaporation

(Effect of concentration on rate of evaporation)

1. Take 6 beakers (150 ml).
2. Clean thoroughly and weigh all the beakers individually and record the values as W_1.
3. Arrange 6 different heating mantles and set at 80 °C.
4. Wait till the mantles acquire the desired temperatures.
5. Prepare different concentrations of NaCl (2%, 4%, 6 %, 8%, 10%, and 20%) in 100 ml.
6. Add 50 ml of solutions in to the weighed beakers.
7. Weigh all the beakers individually and record the weight as W_2 (beaker + solution).
8. Place the beakers on the mantles.
9. Heat the solutions for at least 30 min.
10. After completion of 30 min carefully take out all the beakers.
11. Allow to cool the solutions in the beakers.
12. Weigh all the beakers individually and record the weight as W_3.
13. Find out the weight of water evaporated.
14. Find out the volume of solutions remaining after evaporation with the help of a measuring cylinder (C) ml.
15. Calculate the rate of evaporation in terms of g/min and ml/min.
16. A graph is plotted by taking concentration of solution (%) on X axis and rate of evaporation in ml/min or gm/min on Y axis.
17. Effect of concentration on rate evaporation is compared.

Effect of Viscosity on Evaporation

1. Take 5 beakers (150 ml).
2. Clean thoroughly and weigh all the beakers individually and record the values as W_1.
3. Arrange 5 different heating mantles and set at 80 °C .
4. Prepare different concentrations of glycerin in 100 ml of water (5%, 10%, 20%, 30%, 40%).
5. Weigh all the beakers individually and record the weight as W_2 (beaker + solution).
6. Determine the weight of solution present in the beakers as (A gm) ($W_2 - W_1$).
7. Place the beakers on the mantles.
8. After completion of 30 min carefully take out all the beakers.
9. Allow to cool the solutions in the beakers.
10. Weigh all the beakers individually and record the weight as W_3.
11. Find out the weight of the solution remaining in the beakers.
12. Calculate the weight of water Evaporated.
13. Calculate the rate of Evaporation (gm/min or gm/sec).
14. Find out the volume of solutions remaining after evaporation with the help of a measuring cylinder (C) ml.
15. Find out the Volume of water Evaporated from each beaker and record it as (D) ml.
16. Calculate the rate of evaporation in ml/min or ml/sec.
17. Plot a graph of % of glycerin added on X axis and amount of water evaporated ml/min. or gm/min on Y axis.
18. Compare the effect of viscosity on rate evaporation.

OBSERVATIONS AND CALCULATIONS

Table 45.1: Effect of Temperature on Rate of Evaporation

Sl No	Temperature °C	(W_1) gm	(W_2)gm	Weight of water before evaporation ($W_2 - W_1$) *a* gm	Weight of beaker + water after evaporation (W_3)gm	Weight of water remaining after evaporation($W_3 - W_1$)*b* gm	Weight of water evaporated *a* – *b* (gm)	Rate of evaporation *a* – *b* /30 (gm/ min)	Amount of water remaining after evaporation (C ml)	Amount of water evaporated (D ml)	Rate of evaporation D/30 (ml/min) or 50 – C/30
1	70										
2	80										
3	100										
4	120										
5	140										

Table 45.2: Effect of Surface Area on Rate of Evaporation

Sl No	Volume of the beaker(ml)	Surface area of the beaker (cm^2)	W_1 (gm)	W_2 (gm)	Weight of water before evaporation ($W_2 - W_1$) (*a* gm)	Weight of beaker + water after evaporation (W_3) gm	Weight of water after evaporation ($W_3 - W_1$) *b* (gm)	Weight of water evaporated ($a - b$)gm	Rate of evaporation ($a - b$/30) gm/min	Amount in ml remaining after evaporation (C) ml	Amount of water Evaporated (D ml)	Rate of evaporation D/30 (ml/min)
1	50											
2	100											
3	250											
4	500											

Table 45.3: Effect of Added Substance on Rate of Evaporation

Sl No	Concentration of the solution (%)	Weight of the empty beaker W_1 (gm)	Weight of beaker + solution before evaporation W_2 (gm)	Weight of solution before evaporation ($W_2 - W_1$) *a* gm	Weight of beaker + solution after evaporation W_3 (gm)	Weight of solution after evaporation ($W_3 - W_1$) *b* (gm)	Weight of water evaporated (a – b)(gm)	Rate of evaporation ($a - b$ /30)gm/min	Quantity in ml remaining after evaporation (C ml)	Amount of water Evaporated 50 – C (D ml)	Rate of evaporation D/30 (ml/min)
1	2										
2	4										
3	6										
4	8										
5	10										
6	20										

Table 45.4: Effect of Viscosity on Rate of Evaporation

Sl No	Concentration of Glycerin added (%)	(W_1) gm	(W_2) gm	Weight of water before evaporation ($W_2 - W_1$) *a* gm	Weight of beaker + water after evaporation (W_3) gm	Weight of water remaining after evaporation ($W_3 - W_1$) *b* gm	Weight of water evaporated *a* – *b* (gm)	Rate of evaporation *a* – *b*/30 (gm/ min)	Amount of water remaining after evaporation (C ml)	Amount of water evaporated 50 – C (D ml)	Rate of evaporation D/30 (ml/min)
1	5%										
2	10%										
3	20%										
4	30%										
5	40%										

REPORT

The effect of surface area, temperature, concentration and viscosity on rate of evaporation is studied and the findings are,

The rate of evaporation .... temperature.

The rate of evaporation .... with surface area.

The rate of evaporation .... with concentration.

The rate of evaporation .... with viscosity.

EXPERIMENT 46

EVAPORATION—VERIFICATION OF MATERIAL BALANCING EQUATION

AIM

To verify the material balancing equation during evaporation.

REQUIREMENTS

Beaker	(1)
Water bath	(1)
Tripod stand	(1)
Glass rod	(1)
Measuring cylinder	(1)
China dish	(1)
Hot air oven	(1)
Filter paper	
NaCl	

PRINCIPLE

Material quantities can be expressed in many ways: weight/weight (w/w), weight/volume (w/v), molar concentration (M), mole fraction and weight fraction. The weight/weight concentration is the weight of the solute divided by the total weight of the solution and this is the fractional form of the percentage composition by weight. The weight volume concentration is the weight of solute in the total volume of the solution. The molar concentration is the number of molecular weights of the solute expressed in kg per 1 m^3 of the solution. The mole fraction is the ratio of the number of moles of the solute to the total number of moles of all species present in the solution. The weight fraction is the ratio of weight of solute to the total weight of the species present in the solution.

$$\boxed{\textbf{Mass In = Mass Out + Mass Stored}} \quad \text{... (Eq. 46.1)}$$

Evaporation is a process of removal of liquid from a solution by boiling the liquid in a suitable vessel and withdrawing the vapour leaving a concentrate liquid residue. During this process the mass (material) transfer happens according to the law of conservation. The material balance during evaporation can be determined by the following equation.

$$\boxed{F\, x_F = L\, x_L + V y} \quad \text{... (Eq. 47.2)}$$

where,

F = Initial quantity of solution (liter)
x_F = Weight fraction of the solute in the feed
L = Amount of liquid remained after time intervals (liter)
x_L = Weight fraction of the solute present in L volume of solution
V = Amount of the liquid evaporated (liter)
y = Weight fraction of solute present in the vapour

PROCEDURE

1. Take 100 ml of water in a beaker.
2. Weigh 10 gm of sodium chloride.
3. Add the sodium chloride in 100 ml of water and stir continuously to get a complete solution of 10% sodium chloride solution.
4. Arrange water bath in a tripod stand and heat it till it reaches 90 °C.
5. Always keep the temperature constant for better results.
6. Place the beaker containing 10% sodium chloride solution in the water bath.
7. Note down the time (zero time).
8. Exactly after 10 min transfer the solution in the beaker to a measuring cylinder.
9. Note down the volume of solution in ml.
10. Keep it again in the water bath and heat it for another 10 more min.
11. After 10 min transfer it in to measuring cylinder and find out the volume.
12. This procedure is repeated until the liquid volume reaches ¼th of the original volume (25 ml).
13. When the volume reaches almost ¼th of the total, pour the solution to a pre weighed china dish and heat it to get the semisolid residue in the china dish.
14. Completely dry the residue in the china dish by keeping in it in a hot air oven at 60 °C.
15. Weigh the china dish containing residue.
16. Determine the weight of the residue.
17. Determine the weight fraction of the solute present in the feed (initial solution).
18. Find out the weight fraction of the solute present in the solution (25 ml).
19. Find out the weight fraction of solute present in the vapor.
20. Verify the material balancing equation.

OBSERVATIONS AND CALCULATIONS

Weight of the empty china dish $\mathbf{W_1}$ = ... gm

Amount of solution added to the china dish = 25.......... ml

Weight of the china dish + residue after drying $\mathbf{W_2}$ = ... gm

Weight of the residue $\mathbf{W_2 - W_1}$ = ... gm

Material Balancing Equation

$$F x_F = L x_L + V y$$

F = Initial quantity of solution in liter

= ......0.1 liter.........

x_F = Weight fraction of the solute in the feed

Weight fraction of the solute = $n/n + N$

n = Weight of the solute

= ....kg

N = Amount of solvent in liter

= .... litre

L = The amount of liquid remained after time interval.

= .... litre

x_L = Weight fraction of the solute present in L volume of solution

= ....

V = Amount of the liquid evaporated in litre (F–L)

= .... litre

y = Weight fraction of solute present in the vapor

= $(F_{XF} - L_{XL}/V)$

= ....

REPORT

Verified the material balancing equation of evaporation.

EXPERIMENT 47

EVALUATION OF MATERIALS USED IN PHARMACEUTICAL PACKAGING—CORRUGATED BOARD

AIM

To evaluate the given packaging material (Corrugated board).

REQUIREMENTS

Corrugated board	
(Cut in to pieces of 5 inch × 4 inch)	(2 pieces)
Screw gauge	(1)
Scale	(1)
Beaker	(1)
Tray	(1)
Fractional weight box	(1)

PRINCIPLE

Corrugated board can be defined as the structure formed by gluing one or more fluted corrugated medium to one or more flat facings of liner board. It has been used as a packaging material for more than 100 years and still maintains a healthy share in the market despite increasing competition from plastics. The major advantages of the corrugated boards are:

1. High stiffness to weight ratio
2. Reasonable shock resistance and cushioning properties
3. Fully biodegradable and recyclable
4. It is versatile and can be used for various packaging requirements of different industries

Corrugated board usually consists of two basic components. Line board and corrugating medium. There are mainly four types corrugating medium as follows.

Flute type	Flute spacing (mm)	Flute per meter	Flute height
A	8.3 – 10	110 +/– 10	4.67
C	7.1 – 8.3	130 +/– 10	3.61
B	6.1 – 6.9	165 +/– 10	2.46
E	3.2 – 3.6	290 +/– 15	1.15

The strength of the corrugated board depends upon the number of plies, thickness of the board, number of flutes per unit length of the board, inter spacing between to flutes, and grammage (GSM). The weight per unit area of paper or board is the fundamental material property that must be controlled in the manufacturing stage. Grammage is (also known as the basic weight) defined as the weight of a square meter of paper or board in grams.

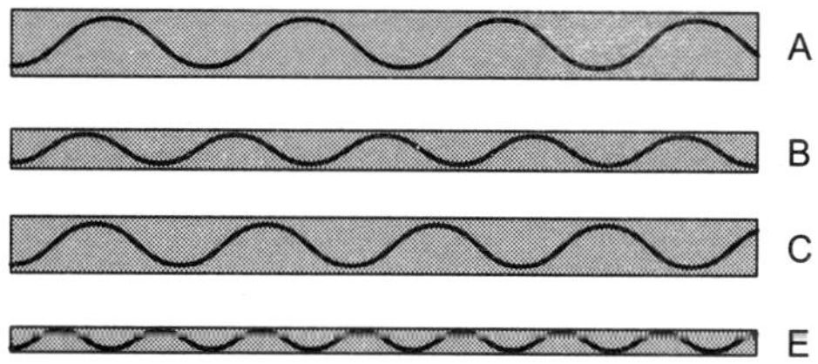

Fig. 47.1: Different flute types.

Common grammage of line board ranges from 125-330 g/m^2 and for corrugated medium 125-250 g/m^2. It is a common practice to match lower grammage liner with lower grammage corrugated medium and higher grammage liner with higher grammage corrugated medium.

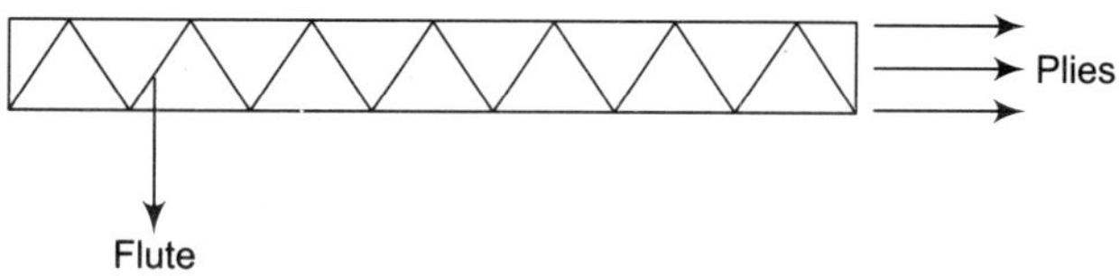

Fig. 47.2: Flutes and plies.

The GSM can be calculated by the equation,

$$\text{GSM} = \frac{\textbf{Weight of piece of paper (gm)} \times \textbf{10000}}{\textbf{Surface area of the piece of paper (cm}^2\textbf{)}} \qquad \text{... (Eq. 47.1)}$$

PROCEDURE

1. **To find out the thickness of the corrugated board:** Place the given corrugated board of known length and breadth between the arms of screw gauge. (Refer to appendix VI). This procedure is repeated thrice and calculates the average thickness.
2. **To find out the number of plies:** The number plies of the board are determined by counting the layers. It include corrugated layers and base layers.
3. **To determine the interspacing between two flutes:** Holding a scale parallel to the corrugated board, the interspacing between the two troughs/peaks of two flutes is determined.
4. **To measure the flute per unit length of the corrugated board:** Holding a scale parallel to the base and corrugated layer the number of flutes per one inch is counted.

5. **To find out the (GSM) Grammage of paper used for making corrugated board:** Take a piece of corrugated board (of known length and breadth). Immerse it into hot boiling water for a period of 5 min (Never more than 5 min). Take out the corrugated board with a holder. Carefully separate the various plies. Transfer the separated piles into a tray containing cold water. After 2-3 min the layers are taken out from the cold water, and remove the adhesive layer by simple rubbing. Dry the layers (air drying) for 10-15 min. Place the layers in a metal tray and place it in a hot air oven for 15 min at 60 °C.Take it out after prescribed time. Trim the sides of all the layers. Find out the area of each layer. Weigh all the plies individually. Note down the readings. Determine the grammage with the help of equation and reported.

OBSERVATIONS AND CALCULATIONS

Thickness of the corrugated board = ... cm

Number of plies = ...

Number of flutes/unit length = ... flutes/ inch

Interspacing between two flutes = ... cm

Flute type = ...

$$\text{GSM} = \frac{\text{Weight of piece of paper (gm)}}{\text{Surface area of the piece of paper (cm}^2)} \times 10000$$

Table 47.1: Determination of Grammage of the Given Corrugated Board

Sl No	*Type of layer*	*Length of the ply (cm)*	*Width (cm)*	*Area of the layer (cm²)*	*Grammage*	*Average grammage of plane layer*	*Average grammage of fluted layer*
1							
2							
3							
4							
5							
6							
7							
8							

REPORT

The given piece of corrugated board is found to have,

Thickness = ... cm

Number of Plies = ...

Flute per inch = ...

Interspacing between two flutes = ... cm

Type of flutes = ... cm

Grammage of plane layer = ... GSM

Grammage of fluted layer = ... GSM

Experiment 48

Evaluation of Packaging Material—Glass

AIM

To determine the hydrolytic resistance of glass container (limit of alkalinity).

REQUIREMENTS

Ampoules	(10)	Methyl red solution
Mortar and pestle	(1)	Acetone
Hot air oven	(1)	H_2SO_4
Autoclave	(1)	NaOH
Sieves	(1)	
Conical flask (250 ml)	(4)	
Chemical balance	(1)	
Spatula	(1)	
Beaker (500 ml)	(1)	
Beaker small	(1)	
Burette	(1)	
Burettes stand	(1)	
White tile	(1)	
Measuring cylinder (100 ml)	(1)	

PRINCIPLE

Glass is commonly used in pharmaceutical packaging because of its excellent protective qualities. It is chemically inert, impermeable, strong, and rigid. Container glasses are made by fusing silica (in the form of sand) with two or other materials, *e.g.*, soda ash and lime stone at very high temperatures. Glass containers for pharmaceutical use are intended to come in to direct contact with pharmaceutical preparations. Glass used for pharmaceutical containers is either a borosilicate glass or a soda lime glass. Borosilicate glass contains boric oxide, aluminum oxide, and alkali and/or alkaline earth oxides. It has a high hydraulic resistance. They are classified as Type 1 glass. On the other hand soda lime glass contains alkali metal oxides and considered to have a moderate hydraulic resistance. This will comes under Type 3 glass. If the soda lime glass is treated to improve the hydrolytic resistance they are categorized under Type 2 glasses. The quality of the glass containers is dependent on their resistance to chemical attack.

One of the major problems with the glass container is the alkali extraction when aqueous preparations are stored for a fairly long period. This may cause the increase in the pH of the solution. Some medicaments are unstable at high pH and some will precipitate during storage. So the alkalinity check is considered as one of the most important and primary tests conducted on glass containers if they are intended to store the liquid preparations.

This can be done by two methods.

1. Crushed glass test
2. Whole container test (mainly for type II glass)

Whole container test is considered as easier and will give more useful information. If the container passes the whole container test it is sure that there will not be any surface extraction of the alkali to the preparation. But by performing this test only one cannot predict the extraction of the alkali from the deeper layers. So it is advisable to perform both the tests to get more accurate results.

PROCEDURE

I. Preparation of Reagents

High purity water : Distilled de ionized water is used.

Methyl Red Solution

Dissolve 24 mg of methyl red solution in purified water to make 100ml. if necessary add 0.02 N sodium hydroxide or 0.02 N sulfuric acid to adjust the pH 5.6.

II. Powdered Glass Test

Select six or more containers (Ampoules/Vials).

1. Rinse thoroughly with purified water.
2. Dry them by passing dry air.
3. Crush the container in to powder with the help of a mortar and pestle (care should be taken to protect eyes and nose while crushing the containers).
4. Transfer in to a sieve assemble (Sieve no 20, 48 and 60).
5. Sieve it for 10 min.
6. Take out the fine powder which retains on the No 60 mesh.
7. The powder which is present over the No 48 and No 20 mesh is transferred in to the mortar and crush it further.

8. This procedure can be repeated till you get 15 gm of fine powder.
9. Transfer the powder in to a 250 ml conical flask.
10. Wash it with six 30 ml portion of acetone; agitate it each time for 30 sec.
11. Carefully decant the acetone (agglomerates of glass are removed if present).
12. Dry the flask and content for 20 min at 140 °C in a hot air oven.
13. Transfer it into a bottle.
14. Weigh carefully 10 gm of powdered sample.
15. Transfer it to a 250 ml conical flask.
16. Add 50 ml of high purity water.
17. Cap the flask by a borosilicate beaker. (Beaker should be selected in such a way that while capping; it should fit on the top rim of the flask.)
18. Place the conical flask in the auto clave, and close it tightly, leaving the vent cock open. Heat till the steam comes out from the vent cock, wait for 10 min and close the vent cock. Adjust the temperature at 121 °C. Wait till the temperature reaches 121 °C. Exactly from this after 30 min, the autoclaving is stopped. Allow to cool the auto clave to the room temperature. Open the vent after five min the lid of the autoclave is opened. And carefully taken out the conical flask. (if the auto clave is pressure regulated adjust the pressure to 15 bars instead of temperature)
19. Cool the flask by keeping under running water.
20. Remove the cap of the conical flask and decant the water to another conical flask.
21. Add 15 ml portion of high purity water to the residual powder and decant in to the main portion (repeat this procedure 4 times). Make up the volume to 100 ml.
22. Add 5 drops of methyl red solution and titrate immediately with 0.02 N sulfuric acid.
23. The volume of 0.02 N sulfuric acid is noted when the color change occurs from yellow to red.
24. A blank determination is also carried out without glass powder.
25. Subtract the blank value from the test value.
26. This result is compared with the standard values as given in Table 48.1.

Table 48.1: Test Limit for Powdered Glass Test

Sl No	*Type of glass*	*Limits*
		ml of 0.02 N sulfuric acid
1	I	1.0
2	III	8.5

III. Test for Water Attack (Whole Container)

1. 5-8 containers (depending upon the volume) are selected for this test.
2. Rinse all the containers thoroughly with purified water.
3. Fill each container with 90% of its over flow capacity with high purified water.
4. Cap the containers with a neutral glass or aluminum foil previously rinsed with purified water. (seal the tip if ampoules are used)
5. Place it in beaker or conical flask in an upright position and place cotton tightly inside the beaker to make the bottles undisturbed.

6. Place the beaker containing glass containers in the auto clave, and close it tightly, leaving the vent cock open. Heat till the steam comes out from the vent cock, wait for 10 min and close the vent cock. adjust the temperature at 121 °C. Wait till the temperature reaches 121 °C. Exactly from this period after 15-20 min, the autoclaving is stopped. Allow to cool the auto clave to the room temperature. Open the vent after five min the lid of the autoclave is opened. And carefully taken out the beaker.
7. Remove the cotton and carefully take out the containers.
8. Remove the caps and transfer the water in to a 100 ml measuring cylinder.
9. If the containers are small, empty the contents from more than one container to make the volume 100 ml.
10. Transfer the contents in to a 250 ml conical flask.
11. Add 5 drops of methyl red solution and titrate immediately with 0.02 N sulphuric acids.
12. The volume of 0.02 N sulfuric acid is noted when the color change occurs from yellow to red.
13. A blank determination is also carried out using the 100 ml of water at the same temperature and same amount of indicator.
14. Subtract the blank value from the original (test) value.
15. This result is compared with the standard values as given in Table 48.2.

Table 48.2: Test Limit for Whole Container Test

Type	*Description*	*Type of test*	*Volume*	*Limits*
				ml of 0.02 N sulfuric acid
II	Treated soda lime glass	Water attack	100 or less	0.7

OBSERVATIONS AND CALCULATIONS

Table 48.3: Powdered Glass Test

Sl No	*Content*	*Initial reading (ml)*	*Final reading (ml)*	*Volume of 0.02 sulfuric acid required for neutralization (ml) final –initial*	*Corrected volume (ml) Sample – blank*	*standard value*	*Pass/ fail the test*
Sample	100 ml of decanted solution and washed extract of crushed glass + 5 drops of methyl red solution						
Blank	100 ml of high purity water + 5 drops of methyl red solution						

Table 48.4: Whole Container Test

Sample	*Contents*	*Initial reading (ml)*	*Final reading (ml)*	*Volume of 0.02 N sulphuric acid used (ml) final – initial*	*Corrected volume Sample – blank (ml)*	*Standard value*	*Pass/ fail test*
Sample	100 ml of water from sample bottles + 5 ml of methyl red solution						
Blank	100 ml of high purified water + 5 ml methyl red solution						

REPORT

The given sets of containers pass/fail the crushed glass (powdered glass) test.

The given sets of containers pass/fail the water attack test.

EXPERIMENT 49

DETERMINATION OF RATE OF CORROSION—TOTAL IMMERSION TEST

AIM

To determine the rate of corrosion of metal pieces in different solutions.

REQUIREMENTS

Metal pieces (Iron) diameter 20 mm and thickness of 3mm	(3)
Beaker	(3)
Round bottom flasks	(3)
Heating mantles	(3)
Glass rods	(3)
Reflux condensers	(3)
Weighing balance	(1)
Weighing Box	(1)
Concentrated solutions of acids	
(1 M H_2SO_4, 1M HCl, 1M NaOH)	

PRINCIPLE

Corrosion can be defined as the degradation of a material due to a reaction with its environment. In broad terms, the interaction between a material and its environment that results in a degradation of the physical, mechanical, or even esthetic properties of that material. More specifically, corrosion is usually associated with a change in the oxidation state of a metal, oxide, or semiconductor. Resistance to a corroding medium is often an important factor in the selection of material for construction of a chemical plant. Laboratory corrosion tests are frequently the quickest and most satisfactory means of selection of most suitable materials for chemical plants. The convectional laboratory type corrosion test gives information about chemical resistance of the proposed materials against dissolved gas, velocity, turbulence, abrasion, crevice condition, hot wall effect, cold wall effect, stress and impurities.

The laboratory test which determines the extend of corrosion is known as total immersion test. In this exporiment, small pieces of metals are exposed to the medium and loss of mass of the metal is measured for a given period of time. It is an unaccelerated method giving reasonable concordant results. During the test the metal pieces are immersed totally in the proposed solutions for a fixed time. And the corrosion produced after the planned interval is noted. The metal pieces are examined for localized pits, crevice corrosion and cracking, If any of this is present the size, shape, and distribution of pits should be noted. The specimens are also subjected to simple bending test to determine whether any embitterment has occurred.

The corrosion rate or penetration can be determined by the equation,

$$\textbf{Corrosion rate (mm/year)} = \frac{\textbf{Weight loss (mg)} \times 13.56}{\textbf{Area (inch}^2) \times \textbf{Time (hr)} \times \textbf{Metal density (g/cm}^3)} \quad \text{... (Eq. 49.1)}$$

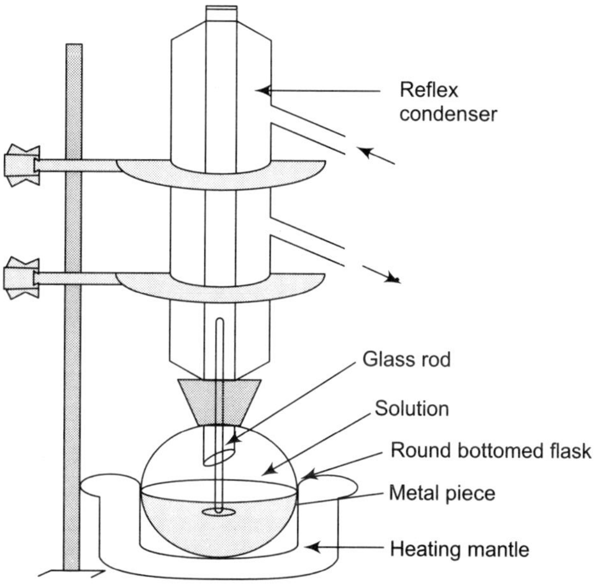

Fig. 49.1: Assemble for the determination of corrosion rate.

PROCEDURE

1. Take 3 circular metal pieces of diameter 20 mm and thickness of 3 mm with a hole at the centre (This hole should be in such a way that a glass rod can tightly fit in to it.) and mark it I-III.
2. Prepare different solution of same strength. (1M H_2SO_4, 1 M NaOH, 1M HCl)
3. Weigh each metal piece and record the initial weights.
4. Place 200 ml of 1M H_2SO_4, 1 M NaOH, 1M HCl in different round bottomed flasks.
5. The temperature of the heating mantles is set to 45 °C.
6. Place the glass rods tightly through the central holes of the metal pieces. (Each piece with one glass rod)

7. Place it in the solutions ie I in 1M H_2SO_4, II in 1 M NaOH and III in 1M HCl (The metal pieces should be completely immersed in the solution).
8. Immediately note down the time. Attach the reflex condensers to the round bottomed flasks. (Refer to Fig. 49.1).
9. Shake the solutions occasionally.
10. Experiments are preceded for 2 hrs.
11. After prescribed time detach the condensers from individual units and takeout the metal pieces.
12. Before cleaning the metal pieces examine it for the development of pits or deposition.
13. Rub the surface of the metal pieces with the help of a mild scrubber to remove the corroded residues.
14. If pitting or crevice or cracking are seen note down the depth and area affected.
15. Now reweigh all the 3 pieces.
16. Find out the differences from the initial weight.
17. The corrosion rate of all the metal pieces is calculated individually with the help of the equation.
18. The data are tabulated and compare the results.

Note: This experiment can be repeated using metal pieces of different materials (Aluminium, Tin, lead etc) for the determination of rate of corrosion.

OBSERVATIONS AND CALCULATIONS

Table 49.1: Local Examination for the Determination of Type of Corrosion

Sl No	*Corroding solution*	*Type of corrosion (localized pits/ crevice corrosion/cracking)*	*Size of the pits*	*Shape of the pits*	*Distribution*	*Bending*
1	1M H_2SO_4					
2	1M HCl					
3	1M NaOH					

Table 49.2: Determination of Rate of Corrosion

Sl No	*Solution used*	*Time of exposure*	*Initial weight (mg) W_1*	*Final weight (mg) W_2*	*Weight loss (mg) $W_1 - W_2$*	*Rate of corrosion (mm/y)*
1	1 M H_2SO_4					
2	1 M HCl					
3	1M NaOH					

Diameter of the metal piece (**D**) = ...mm
= ....inch

Diameter of the mounting hole (***d***) = ...mm
= ....inch

Thickness of the metal piece (***t***) = ...mm
= ....inch

$$\text{Surface area of the metal piece } \textbf{(A)} = \frac{\Pi}{2}\left(D^2 - d^2\right) + t\,\Pi\,D + t\,\Pi\,d$$

$$= \ldots\ldots\ldots\ldots\ldots\ldots \text{ inch}^2$$

$$\text{Metal density} = \ldots\ldots\ldots\ldots\ldots\ldots \text{ g/cm}^3$$

$$\textbf{Corrosion rate (mm/year)} = \frac{\textbf{Weight loss (mg)} \times \textbf{13.56}}{\textbf{Area (inch}^2\textbf{)} \times \textbf{Time (hr)} \times \textbf{Metal density (g/cm}^3\textbf{)}}$$

REPORT

The rate of corrosion is determined for the same material in different solutions and the results are compared.

EXPERIMENT 50

DETERMINATION OF FACTORS AFFECTING CORROSION RATE—TOTAL IMMERSION TEST

AIM

To determine the effect of different factors affecting rate of corrosion.

REQUIREMENTS

Metal pieces (iron) diameter 20 mm and thickness of 3 mm	(12)
Beaker	(6)
Round bottom flasks	(6)
Heating mantles	(6)
Glass rods	(6)
Reflux condensers	(6)
Weighing balance	(1)
Weighing Box	(1)
Concentrated solutions of acids	
(1 M H_2SO_4, 1M HCl, 1M NaOH)	

PRINCIPLE

Refer to previous experiment.

Corrosion of metals will happen at a rate that varies substantially, depending on the conditions. Nature of the material or alloy, surface condition/roughness, material, configuration, material-material spacing, composition, moisture, absorptivity, Structure, Nature of any reinforcement, Temperature, Humidity, Corrosive elements (type, concentration) are some of the factors affecting corrosion rate. Higher levels of corrosion that occur in the presence of moisture and the hostile gas species, such as chlorine, ammonia, sulphur dioxide, hydrogen sulphide and oxides of nitrogen, that are often present in the atmosphere.

Before selecting the material for the chemical plants an extensive accelerated study is conducted to identify the factors affecting the corrosion of materials. In this present experiment we are trying to expose effect of factors like temperature, concentration and presence of oxygen and composition of material, as model tests.

Like most other chemical reactions, corrosion rates increase as temperature increases. Temperature and pressure of the medium govern the solubility of the corrosive species in the fluid, such as oxygen (O), carbon dioxide (CO), chlorides, and hydroxides. Concentration of the corroding medium is also increases the rate of corrosion. Corrosion is also depending upon the materials used. The plated/coated or alloyed materials shows lesser extend of corrosion compared to the pure materials. The presence of oxygen in water or media to which iron (material) is exposed increases the corrosion rate. The reason for this increase is the rapid reaction between oxygen and the polarizing layer of atomic hydrogen absorbed on the oxide layer. Rate of corrosion can be determined using the following equation,

$$\textbf{Corrosion rate (mm/year)} = \frac{\textbf{Weight loss (mg)} \times \textbf{13.56}}{\textbf{Area (inch}^2\textbf{)} \times \textbf{Time (hr)} \times \textbf{metal density (g/cm}^3\textbf{)}} \quad \text{... (Eq. 50.1)}$$

PROCEDURE

Effect of Temperature on Corrosion Rate

1. Take 3 circular metal piece of diameter 20 mm and thickness of 3 mm with a hole at the centre. (This hole should be in such a way that a glass rod can tightly fit in to it.)
2. Prepare solution of 1M HCl.
3. Weigh each metal piece and record the initial weights and mark it I-III.
4. Place 200 ml of 1M HCl in 3 different round bottomed flasks and mark it as 1, 2 & 3.
5. The temperature of the heating mantles is set to 60 °C, 80 °C, 100 °C and mark it as 1, 2 & 3.
6. Keep the round bottom flasks containing 1 M HCl over the corresponding heating mantles.
7. Fit the glass rods tightly through the central holes of the metal pieces. (Each piece with one glass rod).
8. Place metal pieces in the solutions present in the round bottom flasks. (One in each flask) (The metal pieces should be completely immersed in the solution)
9. Immediately note down the time. Attach the reflex condensers to the round bottomed flasks. (Refer to previous experiment Fig. 49.1)
10. Shake the solutions occasionally.
11. Experiments are preceded till 2 hrs.
12. After prescribed time detach the condensers from individual units and takeout the metal pieces.
13. Before cleaning the metal pieces examine for the development of pits or deposition.
14. Rub the surface of the metal pieces with the help of a mild scrubber to remove the corroded residues.
15. If pitting or crevice or cracking are seen note down the depth and area affected.
16. Now reweigh all the 3 pieces.
17. Find out the differences from the initial weights.
18. The corrosion rate of the metal pieces is calculated individually with the help of the equation.
19. The data are tabulated and compare the results.

Determination of Effect of Concentration

Assembles are arranged for 3 different concentration of same solutions.

Determination of Effect of Vacuum (Effect of Oxygen)

Modification of the assemble is done by giving provision for applying vacuum. Three way/Two way round bottom flasks can be used for this purpose. You can also study the effect of different degree of vacuum by increasing/decreasing the vacuum applied.

OBSERVATIONS AND CALCULATIONS

Diameter of the metal piece (**D**) = ...mm

= ....inch

Diameter of the mounting hole (***d***) = ... mm

= ....inch

Thickness of the metal piece (***t***) = ...mm

= ....inch

$$\text{Surface area of the metal piece (A)} = \frac{\Pi}{2}\left(D^2 - d^2\right) + t\,\Pi\,D + t\,\Pi\,d$$

= .... inch^2

Metal density = ... g/cm^3

$$\textbf{Corrosion rate (mm/year)} = \frac{\textbf{Weight loss (mg)} \times 13.56}{\textbf{Area (inch}^2\textbf{)} \times \textbf{Time (hr)} \times \textbf{metal density (g/cm}^3\textbf{)}}$$

Table 50.1: Local Examination for the Determination of Type of Corrosion

Sl No	*Conditions*		*Type of corrosion (localized pits/ crevice corrosion/cracking)*	*Size of the pits*	*Shape of the pits*	*Distribution*	*Bending*
1	Temp °C	60					
		80					
		100					
2	Con (M)	1					
		2					
		5					
3	Vacuum (Pa)						

Table 50.2: Determination of Rate of Corrosion

Sl No	*Conditions*		*Time of exposure (min)*	*Initial weight (mg) W_1*	*Final weight (mg) W_2*	*Weight loss (mg) $W_1 - W_2$*	*Rate of corrosion (mm/y)*
1	Temp °C	60					
		80					
		100					
2	Con. (M)	1					
		2					
		5					
3	Vacuum (Pa)						

REPORT

The factors affecting rate of corrosion are studied and compared.

The rate of corrosion .... as the temperature increases.

The rate of corrosion ....as the concentration of the corroding medium increases.

The rate of corrosion .... as the applied vacuum (decreasing amount of oxygen).

APPENDIX - I

COMMON UNITS AND CONVERSIONS

1 pound mass	= 453.5924 g
1 ton	= 1000 kg
1 foot	= 30.480 cm
1 inch	= 2.54 cm
1 mile	= 1.60935 meters
1 yard	= 0.9144 meters
1 square feet	= 929.0304 cm
1 square inch	= 0.4516 square centimeters
1 square yard	= 0.836127 square meters
1 cubic foot	= 28,316.85 cubic centimeter
1 gallon	= 3.7853 liters
1 gallon	= 231 cubic inches
1 hour	= 60 minutes
1 hour	= 3600 seconds
1 centigrade	= 1.8 Fahrenheit degree
1 pound force	= 444,822.2 dynes
1 pound	= 4.448222 Newton
1 atm	= 760 millimeters of mercury
1 atm	= 1.01325×10^6 dynes/square centimeter
1 atm	= 1.01325×10^5 newtons/square meter
1 Btu (British thermal unit)	= 251.98 calories
1 Btu	= 1054.4 joules
1 Btu	= 0.2930 watt. Hour
1 Btu/ hr ft^2 (Thermal conductivity)	= 0.00413 cal/sec.cm^2
1 Btu/hr (Thermal conductivity)	= 1.728 j/s.m^2
1 btu/hr/ft^2 °F (Heat transfer coefficient)	= 5.678 j/s m^2 °C

Gravitational acceleration g = 9.8066 m/sec^2

Gravitational acceleration g = 32.174 ft/sec^2

1 mm Hg	= 0.133KPa
	= 133.32 Pa
	= 1.333 milli bar
	= 0.0193336 Psi
	= 1 Torr
	= 1.36 cm water
1 Psi	= 68.9475 millibar
	= 6.8475 KPa
	= 51.7148 mm Hg
	= 51.7148 Torr
1 K pa	= 1000 Pa
	= 10 millibar
	= 7.5006 mm Hg
	= 0.1450 Psi
	= 7.500 Torr
1 Pa	= $1N/m^2$

TEMPERATURE CONVERSION FORMULAS

^oF	$= (^oC \times 5/9) + 32$
^oC	$= (^oF - 32)5/9$
^oR	$= {}^oF + 459.69$
^oK	$= {}^oC + 273.15$
^oK	$= {}^oR \times 5/9$

APPENDIX - II

CONVERSION FACTORS

To convert	In to	Multiplied by
atmosphere	N/m^2	1.01325×10^5
atmosphere	bar	1.01295
atmosphere	$Dynes/cm^2$	1.01295×10^6
atmosphere	Pa or N/m^2	1.01295×10^5
atmosphere	Psi or lb/in^2	14.696
atmosphere	Torr or mm Hg	760
bar	atmosphere	0.9872
bar	$Dynes/cm^2$	1×10^6
bar	Pa or N/m^2	1×10^5
bar	Psi or lb/in^2	14.503861
bar	Torr or mm Hg	750.2838
Btu	Cal_{IT}	251.996
Btu	ft-lb	778.17
Btu	Joule	$1.055.06 \times 10^3$
Btu	K.W h	2.9307×10^{-4}
Btu/lb	Cal_{IT}/g	0.55556
$Btu/ft^2.h$	W/m^2	3.1546
$Btu/ft^2.h.^oF$	$W/m^2.^oC$	5.6783
	$Kcal/m^2.h.K$	4.882
Cal_{IT}	Btu	3.9683×10^{-3}
Cal	Joule	4.184
Cm	Inch	0.39370
Cm	ft	0.0328084
Cm^3	ft^3	3.531467×10^{-5}
Cm^3	m^3	10^{-6}
Centipoise	Kg/m.s	1×10^{-3}
Centipoise	lb/ft.h	2.4191
Centipoise	lb/ft.s	6.7197×10^{-4}
Degree	Radian	1.745329×10^{-2}

Dynes	Newton	$1 X10^{-5}$
Dynes/cm^2	atmosphere	9.872×10^{-7}
Dynes/cm^2	bar	1×10^{-6}
Dynes/cm^2	Pa or N/m^2	0.1
Dynes/cm^2	Psi or lb/in^2	1.4508×10^{-5}
Dynes/cm^2	Torr or mm Hg	7.5028×10
ft	m	0.3048
ft.lb	Btu	1.2851×10^{-3}
ft.lb	Joule	1.35582
ft.lb/s	Btu/h	4.6262
ft^2/h	m^2/s	2.581×10^{-5}
ft^3	m^3	0.0283168
Gallon	Cm^3	0.003785
Grams	Pounds	0.0022046
Horse power	Watt	7.46×10^2
Inch	Meter	2.54×10^{-2}
Inch of mercury	N /m^2	3.37685×10^3
Inch of water	N/m^2	2.4884×10^2
Joules	ft/lb	0.73756
Joule	Btu	9.480×10^{-4}
Joules	Cal	0.23389
Joules	KW.h	2.7778×10^{-7}
Kilo calorie $_{IT}$	Joules	4.18674×10^3
Kilo calorie(thermo chemical)	Joules	4.184×10^3
Kilogram	lb	2.20462
KWh	Btu	3,412.1
KW	Horse power	1.3410
Liter	m^3	1×10^{-3}
Liter	ft^3	0.03532
Lb	Kg	0.45359237
Lb/ft^3	Kg/m^3	16.018
Lb/ft^3	g/cm^3	0.016018
Lb f/inch^2	N /m^2	6.89473×10^3
Lb mol/ft^2.h	Kg mol/m^2.s	1.3562×10^{-3}
Lb	Newton	4.448221
Millimeter of mercury	N /cm^2	133.32
Newton	Dynes	1×10^5
Newton	Kg	0.10197
Newton	lb	0.22481
N /m^2	Lb f/inch^2	1.4503×10^{-4}
Pa or N/m^2	Atmosphere	9.869×10^{-6}
Pa or N/m^2	Bar	1×10^{-5}

Pa or N/m^2	$Dynes/cm^2$	10
Pa or N/m^2	Kg/cm^2	1.020×10^{-5}
Pa or N/m^2	Psi or lb/in^2	1.4508×10^{-4}
Pa or N/m^2	Torr or mm Hg	7.5028×10^{-3}
Psi or lb/in^2	atmosphere	0.068046
Psi or lb/in^2	Bar	0.068948
Psi or lb/in^2	$Dynes/cm^2$	6.8948×10^{4}
Psi or lb/in^2	Kg/cm^2	7.0309×10^{-2}
Ton(refrigeration)	Btu /h	12000
Torr	Newton $/m^2$	133.32
Torr or mm Hg	atmosphere	1.3158×10^{-3}
Torr or mm Hg	Bar	1.3328×10^{-3}
Torr or mm Hg	$Dynes/cm^2$	1.3328×10^{3}
Torr or mm Hg	Kg/cm^2	1.3595×10^{-3}
Torr or mm Hg	In. water	0.5353
Torr or mm Hg	mbar	1.3328
Torr or mm Hg	Pa or N/m^2	133.28
Torr or mm Hg	Psi or lb/in^2	1.934×10^{-2}
Watts	Btu/h	3.413
Watts	Joules/s	1
Watts	Kg.m/s	0.10197
Watts.h	Joules	3600
Yards	Meters	0.9144

APPENDIX - III

SATURATED SALT SOLUTION FOR MAINTAINING CONSTANT RELATIVE HUMIDITY CONDITIONS IN DESICCATORS

Saturated salt solution	*Percentage Relative Humidity at Indicated Temperatures*			
	20°C	*25°C*	*30°C*	*37°C*
Lithium bromide	12	11	11	11
Potassium acetate	24	23	23	23
Magnesium chloride	33	33	32	31
Potassium carbonate	44	43	42	41
Magnesium nitrate	53	52	52	51
Sodium nitrate	66	64	63	62
Sodium chloride	76	75	75	75
Potassium bromide	84	83	82	81
Potassium nitrate	94	93	92	91

APPENDIX - IV

DETERMINATION OF DENSITY

Density is the mass of a unit volume of material (weight /ml).

Specific gravity (20° /20°) the ratio of the mass of a given volume of liquid to that of a equal volume of water, both at 20°C .

Relative density is a synonym of specific gravity.

Table 4.1: Density of Water at Different Temperatures

Temperature °C	*Density (g / ml)*
0	0.9987
3.98	1.0000
10	0.99823
15	0.99973
20	0.99823
25	0.99703
30	0.99569
40	0.99224
50	0.98807
60	0.98324
70	0.97781
80	0.97183

PROCEDURE

1. Clean the density bottle with chromic acid or nitric acid.
2. Rinse it with purified water 2–3 times.
3. Dry the bottle by washing it with methanol or acetone.
4. Weigh the empty density bottle with stopper.
5. Weight is noted as W_1.

6. Fill the bottle with purified water and close it with the stopper. (Some water will come out from the bottle. Wipe out with the help of a clean cloth or tissue paper.)
7. Weigh the bottle along with water.
8. The weight is noted as W_2.
9. Pour out the water, and rinse the bottle with the liquid whose density to be determined.
10. Fill the density bottle with the liquid. Close it with the stopper.
11. Weight of the bottle along with the liquid is noted as W_3.
12. Repeat the experiment 2–3 times and take the average values for calculation.

OBSERVATIONS AND CALCULATIONS

Table 4.2: Determination of Density

Sl No	*Weight of the empty bottle (W_1) g*	*Weight of the bottle +water (W_2) g*	*Weight of bottle +liquid (W_3)g*	*Density of the liquid at t °C (g/cm³)*
1.				
2.				
3.				

Average = ...g /cm^3

Weight of empty bottle with stopper = ...W_1 g

Weight of bottle +water = ...W_2 g

Weight of bottle +liquid = ...W_3 g

Weight of water in the density bottle = $W_2 - W_1$ = ...g

Weight of liquid in the density bottle = $W_3 - W_2$ = ...g

$$\text{Density of the liquid} = \frac{\text{weight of the liquid at t°C } (W_3 - W_1)}{\text{weight of water at t°C } (W_2 - W_1)} \times \text{Density of water}$$

APPENDIX - V

VERNIER CALIPERS

The precision of length measurements may be increased by using a device that uses a sliding vernier scale. This instrument has a main scale (in millimeters) and a sliding vernier scale. The vernier scale is divided into 10 or 50 equal divisions having a least count of 0.1mm or 0.02 mm respectively. Both the main scale and the vernier scale readings are taken into account while making a measurement. The part and uses of the parts are mentioned below.

Outside jaws: Used to measure external diameter or width of an object

Inside jaws: Used to measure internal diameter of an object

Main scale: Gives measurements of up to one decimal place (in mm).

Main scale: Gives measurements in fraction (in inch)

Vernier gives measurements up to two decimal places (in mm)

Vernier gives measurements in fraction (in inch)

Retainer: used to block movable part

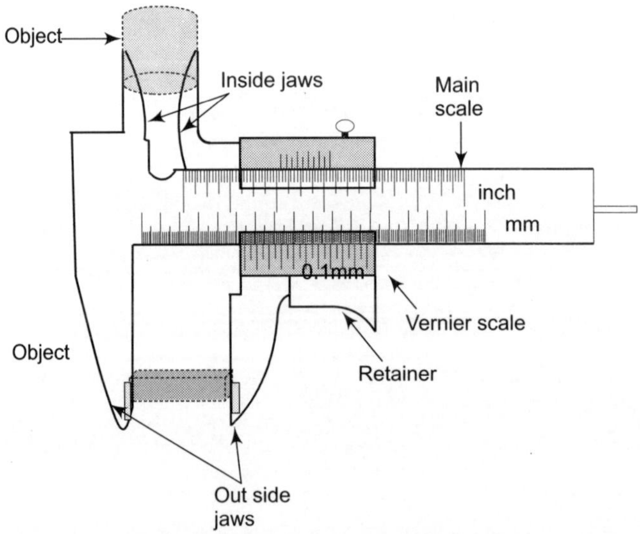

Fig. 5.1: Vernier Calipers.

PROCEDURE

When a caliper's jaws are fully closed, it should of course indicate zero, if not it should recalibrated or repaired.

1. Place the objects as shown in Fig. 5.1.
2. The jaws are then moved together until they secure the object.
3. The screw clamp may then be tightened to ensure that the reading does not change while the scale is being read.
4. The first significant figures are read immediately to the left of the zero of the vernier scale and the remaining digits are taken as the vernier scale division that lines up with any main scale division.

EXAMPLE

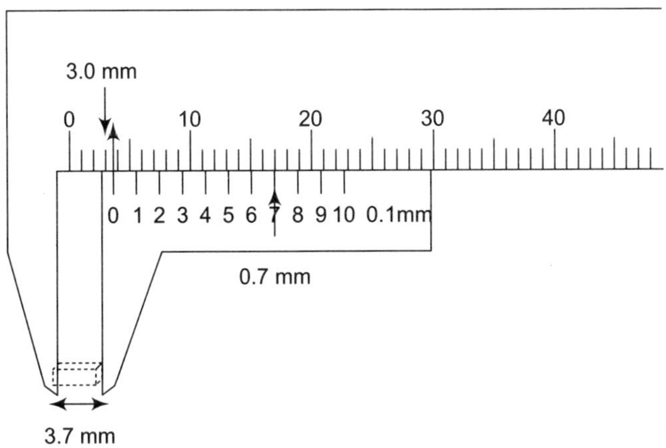

Fig. 5.2: Enlarged portion of the scale.

Main scale divisions present left of the 0 of the vernier scale	= 3 mm
The vernier scale reading that exactly lines up with the main scale reading	= 0.7 mm
The indent length of the object is	= 3.7 mm

APPENDIX - VI

SCREW GAUGE (MICROMETER)

Description

Screw Gauge consists of U shaped metallic frame. To one side of this U frame a long hallow cylindrical tube (Sleeve) with a nut inside is attached, the inner side of cylindrical nut contains a uniform thread cut in it (index line). On the other side of U frame a fixed stud S1 with a plane face is attached.

A screw S2 (Spindle) is fitted in the cylindrical nut. One side of the screw S2 has a plane face similar to that of stud S1 (Anvil). The faces of S1 and S2 are plane and parallel to one another. The other end of the screw S2 carries a milled head 'H' (Thimble) attached to a cap 'C'with a sloping edge. When the head H is rotated, the screw moves "to and fro" in the nut. The milled head H is provided with a safety device 'D' (Ratchet Knob) to rotate the head H. When the object is held between the stud S1 and screw S2 and the head H is rotated using the safety device (D), it Will Produce crackling sound when optimum pressure is applied on the object.

The outer surface of long cylindrical nut consists of a thick horizontal line 'P' parallel to the axis of cylindrical tube. This line 'P' is called Index line. Along the index line a scale is graduated in millimeters. This scale is called Pitch Scale. On the sloping edge of the cap 'C' a circular scale is graduated, which consists of 100 equal divisions, this scale is called Head scale.

The screw gauge works on the principle of a screw. When we rotate the head 'H' by means of safety device 'D' through one complete rotation, the distance moved by the screw for every complete rotation is a constant. This constant distance moved by the screw for one complete rotation of head H is called Pitch of the screw. If the head scale has 100 equal divisions, then the distance moved by the screw for even 1/100 of a complete rotation can be measured accurately, and this is called least count of screw gauge.

Reading = Linear scale reading + (Coinciding circular scale × Least count)

Reading = Pitch scale reading (PSR) +Head scale reading (HSR) × Least count

Least count (L.C) = (Pitch of the screw)/(Number of divisions on Head scale) mm

Pitch of the Screw = (Distance moved by sloped edge over the pitch scale)/(Number of rotations of the screw) mm

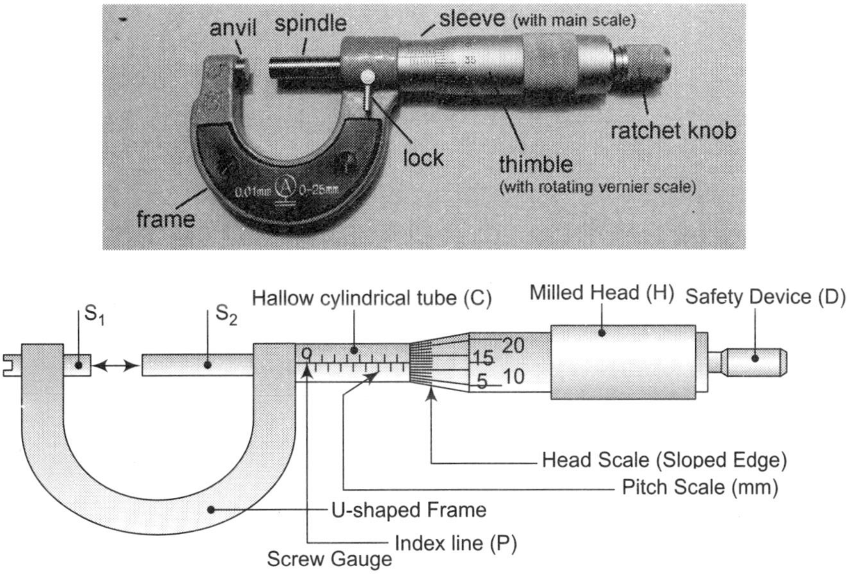

Fig. 6.1: Screw gauge.

PROCEDURE

1. First we have to determine the least count of the given Screw gauge.
2. To determine the least count of the screw gauge, the head 'H' is rotated through certain (say 5) number of complete rotations. The distance moved by the sloped edge over the pitch scale is measured.
3. Now substitute these values in the formula of pitch of the screw = {Distance moved by sloped edge over the pitch scale}/ (Number of rotations of the screw).
4. Least count L.C = {Pitch of the screw}/ (Number of divisions on Head scale).
5. Now check whether the given screws gauge has any zero error or not. To determine the zero error, the head H is rotated until the flat end of the screw S2 touches the plane surface of the stud S1 (do not apply excess pressure) *i.e.* we have to rotate the head only by means of safety device 'D'.

ZERO-ERROR

When S1 and S2 are in contact, the zero of the head scale perfectly coincides with the index line as in Fig. 6.2(*a*). In such case there will be no zero error and no correction is required.

When S1 and S2 are in contact, the zero of the head scale is below the index line as in Fig. 6.2 (*b*), such zero error is called positive zero error, and the correction is negative.

When S1 and S2 are in contact, the zero of the head scale is above the index line as in Fig. 6.2(*c*), such zero error is called negative zero error, and the correction is positive.

Zero Correction

(*a*) No correction

(*b*) – 2

(*c*) + 4

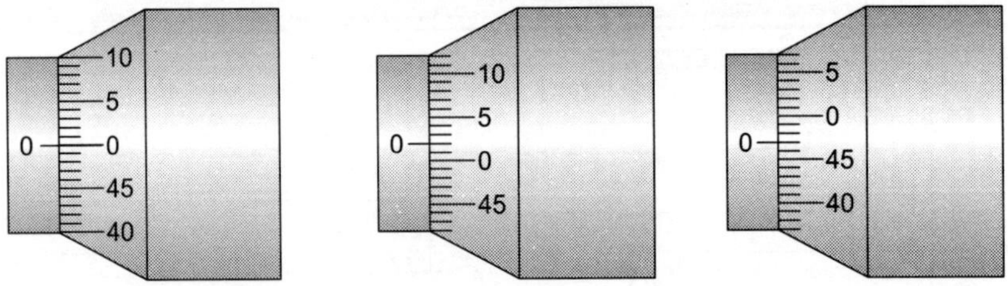

(a) (b) (c)

No zero error Positive zero error Negative zero error

Fig. 6.2: Determination of the radius(r) of the given metal wire.

Determination of Radius of given Metal Wire

1. The given object metal wire is held between the two parallel surfaces of fix stud S1 and screw tip S2 tighen the screw (D) till the cracking sound is obtained.
2. Note the completed number of divisions on pitch scale, which is called pitch scale reading (P.S.R).
3. The number of the head scale division coinciding with the index line is noted, which is called observed head scale reading n'.
4. If the given screw gauge has zero error (x) the correction is made by adding or subtracting the zero error (x) from the observed head scale reading n'. The corrected value $(n' - x)$ or $(n' + x)$ is called the head scale reading (H.S.R) n.
5. Find out the least count of the scale.
6. Find out the Diameter of the wire by the equation.
7. Changing the position of metal wire, 5 readings should be taken, and record in the table.
8. Every time calculate the total diameter (d) of the metal wire using equation.
9. Average of the 5 diameter of the metal wire should be calculated, to get the average diameter (d) of the given metal wire.

PRECAUTIONS

(*i*) Screw must be rotated by holding the safety device 'D'.

(*ii*) Do not apply excess pressure on the object held between the surfaces S1 and S2.

(*iii*) The screw is rotated in only one direction either clock wise or anti-clock wise to avoid the back lash error.

OBSERVATIONS AND CALCULATIONS

Zero error = ...

Zero correction = ...mm

Distance moved by the head for 5 complete revolutions = ... mm

Number of head scale divisions	= ………
Pitch of the screw	= (Distance moved by sloped edge over the pitch scale)/(Number of rotations of the screw).
Least count (L.C)	= (Pitch of the screw)/(Number of divisions on Head scale)

Pitch scale Reading	= …………mm
Head scale Reading (n')	= …………
Zero correction (x)	= ….………
Head scale reading (n)	= ……….
Diameter of the wire	= Pitch scale reading (PSR) + Head scale reading (HSR) × Least count
	= ………… mm

Table 6.1: Determination of Diameter of the Given Wire

Sl. No.	*Pitch scale reading(mm)*	*Head scale reading(n) mm*	*Least count (mm)*	*Diameter of the wire (mm)*	*Average diameter of the wire (mm)*
1					
2					
3					
4					
5					

RESULT

The diameter of the given wire = ………….. mm

APPENDIX - VII

TABLE OF WET AND DRY BULB THERMOMETER READINGS WITH CORRESPONDING PERCENTAGE HUMIDITY

Dry bulb temperature	*Depression of wet bulb temperature °F*													
	1	*2*	*3*	*4*	*5*	*6*	*7*	*8*	*9*	*10*	*11*	*12*	*13*	*14*
30	88	77	66	54	43	33	22	12	1					
32	89	79	68	57	47	36	27	17	7					
34	90	79	70	60	50	41	31	21	13	3				
36	90	80	70	60	54	44	35	26	18	9				
38	91	81	72	63	54	44	39	31	22	14	6			
40	91	82	73	65	56	47	39	30	27	19	11	3		
42	91	83	74	66	58	50	42	34	26	18	16	9		
44	92	84	75	68	60	52	45	37	29	22	15	8	7	
46	92	84	77	69	62	54	47	40	33	26	19	12	6	
48	92	85	77	70	63	56	49	42	36	29	22	16	10	4
50	93	86	79	72	65	59	52	45	38	32	26	20	14	8
52	93	86	79	73	66	60	54	47	41	35	29	23	17	12
54	93	87	80	74	68	61	55	49	43	38	32	26	21	15
56	94	87	81	75	69	63	57	51	46	40	35	29	24	19
58	94	88	82	76	70	64	59	53	48	42	37	31	26	22
60	94	88	82	77	71	65	60	55	50	44	39	34	29	25
62	94	88	83	77	72	67	61	56	51	46	41	37	32	27
64	94	89	83	78	73	68	63	58	53	48	43	39	34	30
66	95	89	84	79	74	69	64	59	54	50	45	41	36	32
68	95	90	84	79	75	70	65	60	56	51	47	43	38	34
70	95	90	85	80	75	71	66	62	57	53	49	44	40	36
72	95	90	85	80	76	71	67	63	58	54	50	46	42	38
74	95	90	86	81	77	72	68	64	60	56	52	48	44	40
76	95	91	86	82	78	73	69	65	61	57	53	49	45	42
78	95	91	86	82	78	74	70	66	62	58	54	50	47	43
80	96	91	87	83	79	74	70	66	63	59	55	52	48	45

82	96	91	87	83	79	75	71	67	64	60	56	53	49	46
84	96	92	87	83	80	76	72	68	64	61	57	54	51	47
86	96	92	88	84	80	76	72	69	65	62	58	55	52	49
88	96	92	88	84	81	77	73	70	66	63	59	56	53	50
90	96	92	88	84	81	77	74	71	67	63	60	57	54	51
92	96	92	88	85	82	78	74	71	67	64	61	58	55	52
94	96	93	89	85	82	78	75	72	68	65	62	59	56	53
96	96	93	89	85	82	78	75	72	68	65	62	59	57	54
98	96	93	89	86	83	79	76	73	69	66	63	60	57	54
100	96	93	89	86	83	79	76	73	70	67	64	61	58	55
102	96	93	90	86	84	80	77	74	70	67	65	62	59	56
104	96	93	90	86	84	80	77	74	71	68	65	62	60	57
106	96	93	90	87	84	80	77	75	71	68	66	63	60	58
108	97	93	90	87	84	81	78	75	72	69	66	63	61	58
110	97	93	90	87	84	81	78	75	72	69	67	64	61	59
112	97	94	90	87	84	81	78	75	73	70	67	65	62	59
114	98	94	90	87	84	81	79	76	73	70	68	65	63	60
116	97	94	90	87	84	82	79	76	73	71	68	65	63	61
118	97	94	91	88	85	82	79	76	74	71	69	66	63	61
120	97	94	91	88	85	82	79	77	74	72	69	67	64	62
125	97	94	91	88	85	83	80	78	75	72	70	68	65	63
130	97	94	91	89	86	83	81	78	76	73	71	69	66	64
135	97	95	92	89	86	84	81	79	76	74	72	69	67	65
140	97	95	92	89	87	84	82	79	77	75	73	70	68	66

APPENDIX - VIII

THE TYLER STANDARD SCREEN SCALE

Mesh	*Clear opening, inch*	*Clear opening mm*	*Opening, inch (approx)*	*Diameter of wire, inch*
....	1.050	26.67	1	0.148
.....*	0.883	22.43	7/8	0.135
.....	0.742	18.85	¾	0.135
.....*	0.624	15.85	5/8	0.120
....	0.525	13.33	1/2	0.105
.....*	0.441	11.20	7/16	0.105
.....	0.371	9.423	3/8	0.092
2 ½ *	0.312	7.925	5/16	0.088
3	0.263	6.680	¼	0.070
3 ½*	0.221	5.613	7/32	0.065
4	0.185	4.699	3/16	0.065
5	0.156	3.962	5/32	0.044
6	0.131	3.327	1/8	0.036
7*	0.110	2.794	7/64	0.0328
8	0.093	2.362	3/32	0.032
9*	0.078	1.981	5/64	0.033
10	0.065	1.651	1/16	0.035
12*	0.055	1.397	....	0.028
14	0.046	1.168	3/64	0.025
16*	0.0390	0.991	.....	0.0235
20	0.0328	0.833	1/32	0.172
24*	0.0276	0.701	....	0.0141
28	0.0232	0.589	....	0.0125
32*	0.0195	0.495	....	0.0118
35	0.0164	0.417	1/64	0.0122
42*	0.0138	0.351	....	0.0100
48	0.0116	0.295	....	0.0092

60*	0.0097	0.246	....	0.0070
65	0.0082	0.208	....	0.0072
80*	0.0069	0.175	....	0.0056
100	0.0058	0.147	....	0.0042
115*	0.0049	0.124	....	0.0038
150	0.0041	0.104	....	0.0026
170*	0.0035	0.088	....	0.0024
200	0.0029	0.074	....	0.0021

*These screens, for closer sizing, are inserted between the sizes usually considered as standard sieves. With the inclusion of these screens the ratio of diameters of openings in two successive screens is as 1: $\sqrt[4]{2}$ instead of 1: $\sqrt{2}$.

APPENDIX - IX

PHOTO COLORIMETER PRINCIPLE

Beer-Lamberts Law

The fraction of incident radiation absorbed is a function of the thickness (path length) and the concentration of absorbing molecules in its path. It can also be represented by the following equation:

$$A = abc$$

When the concentration (c) is expressed in moles/liter the expression becomes:

$$A = \varepsilon b c$$

where, ε = The molar absorptivity

c = The concentration (moles/liter)

b = The length of the light path

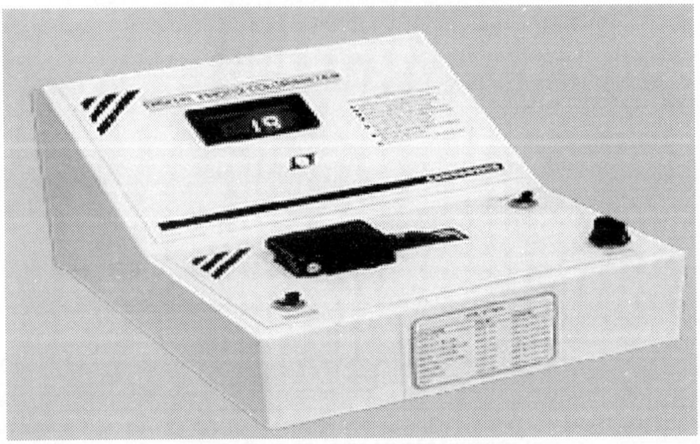

Fig. 9.1: Photo colorimeter.

Determination of amount of salicylic acid present in the sample by spectrophotometer

The sample solution is mixed with specific amount of Ferric chloride or ferric nitrate. The bluish purple complex formed between Fe_3+ ion and salicylic acid allows its visual detection. Obviously, the intensity of the observed color is proportional to the complex concentration. Complex concentration can be determined by spectrophotometer.

Measuring the absorbance of the complex in the visible region (at 540 nm). The Beer—Lambert law states that the proportion of light absorbed by a solute in a transparent solvent is proportional to the number of absorbing molecules in the light path.

$$A = Log\ 1_0\ (I_0/I)$$

$$A = \varepsilon\, l\, C$$

where, I_0 = Intensity of incident light

I = Intensity of transmitted light

ε = Molar absorptivity

C = Sample concentration (in mol/L)

l = Cell (path) length (cm); $b = 1$ cm

A = Absorbance

Therefore, the absorbance is in direct relation with the concentration of the absorbent Species. The concentration can be determined by measuring the absorbance of the sample. Several solutions of known molar concentration must be prepared, and their absorbance measured at the desired wavelength. A plot of absorbance versus concentration gives, according to eq. 1, points that can be joined by a straight line. These plots are called standard curves. By addition of $FeCl_3$ to a sample salicylic acid, a solution of iron-phenol complex with Fe (III) complex is formed. Absorbance of the solution containing the complex is determined. The concentration of the complex will be calculated from the standard curve.

OPERATING PROCEDURE

1. Allow 5 min warm up period after switching ON the instrument.
2. Select the appropriate filter on the filter disc depending upon the test to be performed.
3. Take a blank solution *e.g.*, (distilled water) in the cuvette (up to the mark) and place it in its position in the holder and cover it. Make sure that the mark on the cuvette coincides with that on the panel.
4. Switch ON the shutter control.
5. Adjust the display to ZERO absorbance (optical density).
6. Switch OFF the shutter control.
7. Now remove the blank solution cuvette from the holder.
8. Place the cuvette containing the standard test solution in the holder. Switch ON the shutter. Record the reading (I_s) of absorbance (O.D) on the display.
9. Replace this cuvette with another one containing the solution to be tested and place it in the holder.
10. Record the absorbance (O.D) on the display (I_t).
11. Concentration of unknown solution can be calculated as follows.
    Concentration of unknown solution = I_t/I_s X Concentration of standard solution

PREPARATION OF CALIBRATION CURVE OF SALICYLIC ACID

Stock solution of salicylic acid (0.4 mg/ml)

Accurately weigh about 100 mg of salicylic acid and transfer in to a 250 ml volumetric flask. 3-5 drops of alcohol can be added to dissolve the drug. Distilled water is added in parts and shaken thoroughly after each addition. After ensuring the complete dissolution of drug the volume is made up to 250 ml.

Ferric chloride solution (4%)

Accurately weigh 4 gm of ferric chloride and transfer in to 100 ml amber colored volumetric flask. Add distilled water in parts and shaken thoroughly after each addition. After ensuring the complete dissolution of ferric chloride the volume is made up to 100 ml.

Preparation of standard solutions

1. Dilutions are prepared in 25 ml volumetric flask as shown in the table below.

Sl No	*Standard stock solution (0.4 mg/ml)*	*Volume of ferric chloride solution (4% w/v)*	*Volume of distilled water*	*Concentration of the solution mg/ml*	*Absorbance at 547 nm I_s*
1	1ml	5ml	14 ml	0.02	
2	2ml	5ml	13ml	0.04	
3	3ml	5ml	12ml	0.06	
4	4ml	5ml	11ml	0.08	
5	5ml	5ml	10ml	0.1	
6	0 ml	5 ml	15 ml	.....	

2. The absorbance is determined as described in the operation procedure (The blank solution is the solution containing no drug).
3. A graph is plotted taking concentration in mg/ml on X xis and Absorbance at 547 nm at Y axis.

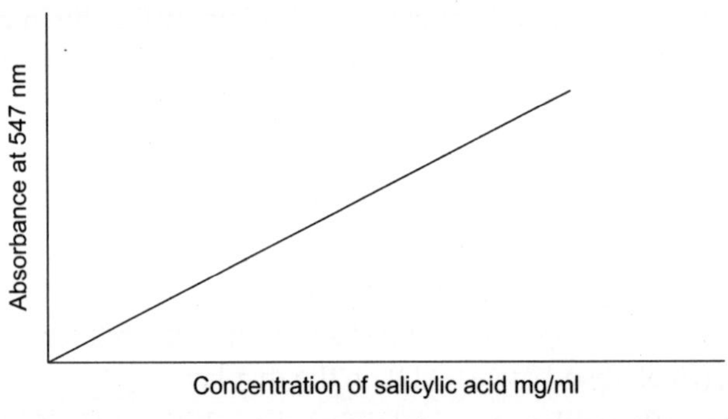

Fig. 9.2: Model standard graph of salicylic acid.

APPENDIX - X

ALCOHOLIMETRIC TABLE

% V/V	ρ_{20} (Kg/m³)	% V/V	ρ_{20} (Kg/m³)	% V/V	ρ_{20} (Kg/m³)	% V/V	ρ_{20} (Kg/m³)
0.00	998.20	25.5	967.53	50.5	929.16	76.0	870.15
0.5	997.44	26.0	966.97	51.0	928.16	76.5	868.82
1.0	996.70	26.5	966.39	51.5	927.16	77.0	867.48
1.5	995.96	27.0	965.81	52.0	926.16	77.5	866.13
2.0	995.23	27.5	965.23	53.0	924.12	78.0	864.78
2.5	994.51	28.0	964.64	53.5	923.09	78.5	863.41
3.0	993.81	28.5	964.04	54.0	922.06	79.0	862.04
3.5	993.11	29.0	963.44	54.5	921.01	79.5	860.66
4.0	992.41	29.5	962.83	55.0	919.96	80.0	859.27
4.5	991.73	30.0	962.21	55.5	918.91	80.5	857.87
5.0	991.06	30.5	961.59	56.0	917.84	81.0	856.46
5.5	990.39	31.0	960.95	56.5	916.77	81.5	855.05
6.0	989.73	31.5	960.31	57.0	915.70	82.0	853.62
6.5	989.34	32.0	959.66	57.5	914.62	82.5	852.19
7.0	988.43	32.5	959.01	58.0	913.53	83.0	850.74
7.5	987.79	33.0	958.34	58.5	912.43	83.5	849.29
8.0	987.16	33.5	957.66	59.0	911.33	84.0	847.82
8.5	986.54	34.0	956.98	59.5	910.23	84.5	846.34
9.0	985.92	34.5	956.29	60.0	909.11	85.0	844.85
10	984.71	35.0	955.59	60.5	908.00	85.5	843.35
10.5	984.11	35.5	954.88	61.0	906.87	86.0	841.84
11	983.82	36.0	954.15	61.5	905.74	86.5	840.31
11.5	982.94	36.5	953.42	62.0	904.60	87.0	838.77
12	982.35	37.0	952.69	62.5	903.46	87.5	837.21
12.5	981.78	37.5	951.94	63.0	902.31	88.0	835.64
13	981.21	38.0	951.18	63.5	901.15	88.5	834.05
13.5	980.64	38.5	950.41	64.0	899.99	89.0	832.45

14	980.08	39.0	949.63	64.5	898.83	89.5	830.82
14.5	979.52	39.5	948.84	65.0	897.65	90.0	829.18
15	978.79	40.0	948.05	65.5	896.47	90.5	827.52
15.5	978.42	40.5	947.24	66.0	895.28	91.0	825.83
16	977.87	41.0	946.42	66.5	894.09	91.5	824.13
16.5	977.33	41.5	945.59	67.0	829.89	92.0	822.39
17	976.79	42.0	944.76	67.5	891.69	92.5	820.63
17.5	976.25	42.5	943.91	68.0	890.48	93.0	818.65
18	975.71	43.0	943.06	68.5	889.26	93.5	817.03
18.5	975.17	43.5	942.19	69.0	888.03	94.0	815.18
19	974.63	44.0	941.32	69.5	886.80	94.5	813.30
19.5	974.09	44.5	940.43	70.0	885.56	95.0	811.38
20	973.56	45.0	939.54	70.5	884.32	95.5	809.42
20.5	973.02	45.5	938.64	71.0	883.06	96.0	807.42
21	972	46.0	937.73	71.5	881.81	96.5	805.37
21.5	17.46	46.5	936.81	72.0	880.54	97.0	803.27
22	17.87	47.0	935.88	72.5	879.27	97.5	801.12
22.5	970.86	47.5	934.95	73.0	877.99	98.0	798.90
23.0	970.31	48.0	934.00	73.5	876.70	98.5	796.62
23.5	969.76	48.5	933.05	74.0	875.40	99.0	794.25
24.0	969.21	49.0	932.09	74.5	874.10	99.5	791.80
24.5	968.66	49.5	931.12	75.0	872.79	100.0	789.24
25.0	968.10	50.0	930.14	75.5	871.48		

APPENDIX - XI

TABLES AND GRAPHS SHOWING VAPOUR PRESSURE OF WATER, METHANOL AND ETHANOL AT DIFFERENT TEMPERATURES

VAPOUR PRESSURE OF WATER (KPa) AT DIFFERENT TEMPERATURES

Tem. °C	0	2	4	6	8	10	12	14	16	18
	PRESSURE IN KPa									
0	0.6112	0.7060	0.8135	0.9353	1.0729	1.2281	1.4027	1.5989	1.8187	2.0646
20	2.3392	2.6452	2.9857	3.3638	3.37809	4.2452	4.7582	5.3240	5.9472	6.6324
40	7.384	8.2096	9.1126	10.100	11.117	12.352	13.632	15.023	16.534	18.173
60	19.948	21.869	23.946	26.188	28.605	31.210	34.013	37.010	40.240	43.704
80	47.416	51.388	55.636	60.174	65.018	70.183	75.685	81.542	87.771	94.390
100	101.42	108.87	116.78	125.15	134.01	143.38	153.28	163.74	174.77	186.41
120	198.67	211.59	225.18	239.47	254.49	270.28	286.84	304.22	322.44	341.54
140	361.53	382.46	404.36	427.25	451.17	476.16	502.24	529.45	557.83	587.41
160	618.23	650.32	683.73	718.48	754.62	792.12	831.22	871.75	913.84	957.51
180	1002.8	1049.8	1098.5	1148.9	1201.2	1255.2	1311.2	1369.1	1429.0	1490.0
200	1554.9	1621.0	1689.3	1759.8	1832.6	1907.7	1985.1	2065.0	2147.3	2232.2

VAPOUR PRESSURE OF ETHANOL (mmHg) AT DIFFERENT TEMPERATURES

Tem °C	0	2	4	6	8	10	12	14	16	18
	PRESSURE IN mmHg									
0	11.77	13.56	15.6	17.89	20.47	23.37	26.62	30.26	34.32	38.85
20	43.89	49.48	55.68	62.54	70.12	78.47	87.67	97.78	108.87	121.03
40	134.33	148.86	164.72	181.99	200.78	221.19	243.35	267.36	293.34	321.43
60	351.71	384.49	419.73	457.66	498.43	542.20	589.16	639.48	693.34	750.94
80	812.48	878.17	948.22	1022.82	1102.30	1186.79	1276.58	1371.91	1473.04	1580.25
100	1693.80	1813.98	1941.07	2075.38	2217.20	2366.86	2524.68	2690.97	2866.09	3050.37

VAPOUR PRESSURE OF METHANOL (mmHg) AT DIFFERENT TEMPERATURES

Tem °C	0	2	4	6	8	10	12	14	16	18
	PRESSURE IN mmHg									
0	30.34	34.30	38.81	43.83	49.41	55.59	62.42	69.96	78.27	87.42
20	97.48	108.51	120.59	133.80	148.23	163.97	181.11	199.75	220.00	241.96
40	265.75	291.49	319.31	349.34	381.72	416.58	454.09	494.39	537.65	584.03
60	633.72	686.89	743.73	804.44	869.21	938.26	1011.81	1090.06	1173.27	1261.65
80	1355.47	1454.95	1560.38	1672.00	1790.10	1914.96	2046.86	2186.10	2332.98	2487.81
100	2650.91	2822.60	3003.22	3193.10	3392.59	3602.05	3821.84	4052.32	4293.88	4546.88

GRAPH SHOWING RELATIONSHIP BETWEEN VAPOUR PRESSURE AND TEMPERATURE OF WATER

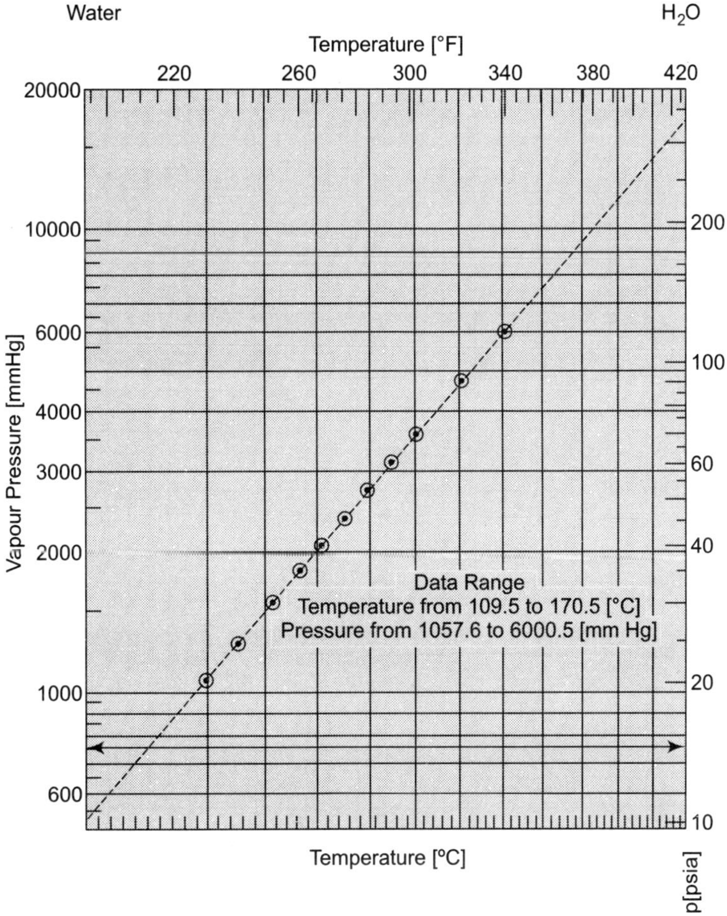

GRAPH SHOWING RELATIONSHIP BETWEEN VAPOUR PRESSURE AND TEMPERATURE OF METHANOL

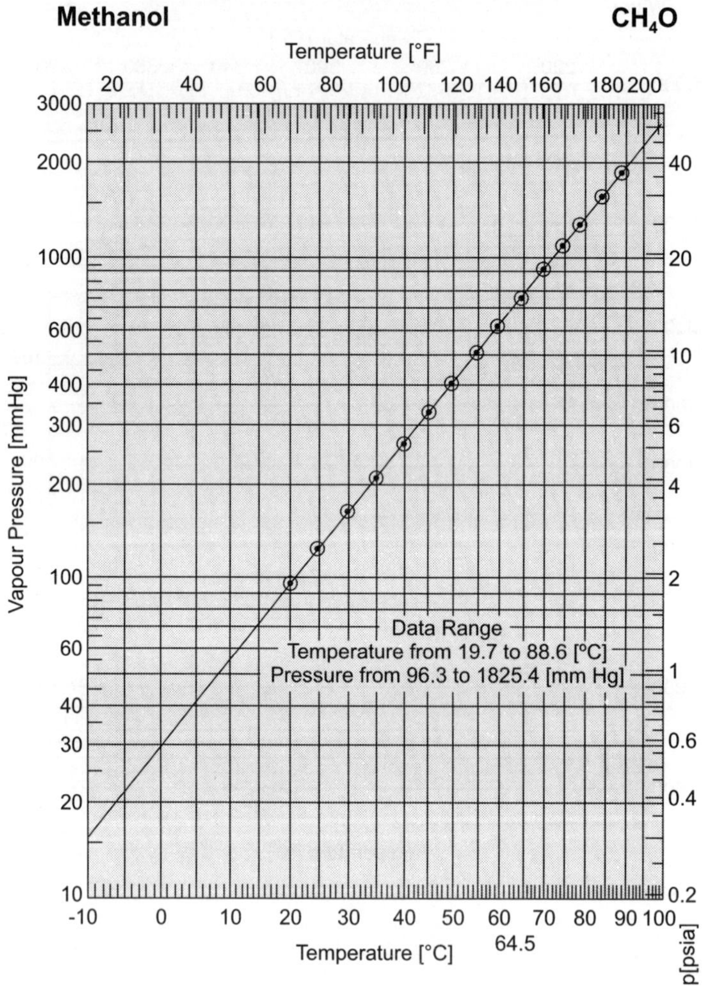

GRAPH SHOWING RELATIONSHIP BETWEEN VAPOUR PRESSURE AND TEMPERATURE OF ETHANOL

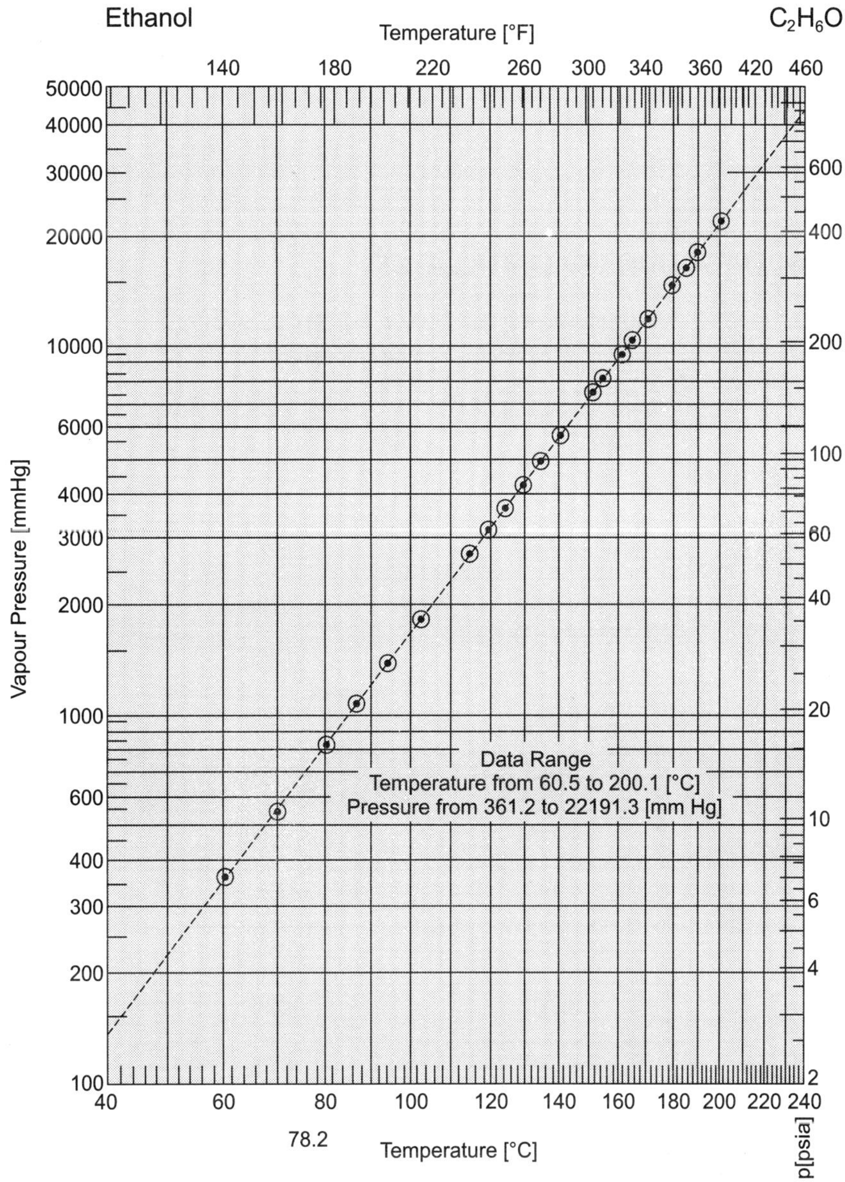

APPENDIX-XII

EQUILIBRIUM DATA DURING DISTILLATION

EQUILIBRIUM DATA OF METHANOL
(Methanol- Water system)

x	y
0	0
0.032	0.190
0.052	0.294
0.059	0.308
0.075	0.352
0.088	0.390
0.154	0.490
0.182	0.552
0.225	0.593
0.290	0.645
0.349	0.703
0.590	0.830
0.813	0.918
0.918	0.963
100	100

EQUILIBRIUM DATA OF ETHANOL

(Ethanol- Water system)

x	*y*
0.019	0.170
0.072	0.389
0.097	0.438
0.124	0.470
0.166	0.509
0.234	0.545
0.261	0.558
0.327	0.583
0.397	0.612
0.508	0.656
0.520	0.660
0.573	0.684
0.676	0.739
0.747	0.782
0.894	0.894

APPENDIX-XIII

SPECIFIC HEAT OF DIFFERENT METALS

Metal	Specific Heat Capacity - c_p		
	(KJ/kg K)	(Kcal/kg°C)	(Btu/lb°F)
Aluminum	0.91	0.22	0.22
Antimony	0.21	0.05	0.05
Beryllium	1.83	0.436	0.436
Bismuth	0.13	0.03	0.03
Cadmium	0.23	0.055	0.055
Carbon Steel	0.49	0.12	0.12
Cast Iron	0.46	0.11	0.11
Chromium	0.46	0.11	0.11
Cobalt	0.42	0.1	0.1
Copper	0.39	0.092	0.09
Gold	0.13	0.031	0.03
Iridium	0.13	0.031	0.31
Iron	0.46	0.108	0.11
Lead	0.13	0.031	0.03
Magnesium	1.05	0.243	0.25
Manganese	0.48	0.114	0.114
Mercury	0.14	0.033	0.03
Molybdenum	0.25	0.06	0.06
Nickel	0.54	0.106	0.13

Niobium(Columbium)	0.27	0.064	0.064
Osmium	0.13	0.031	0.031
Platinum	0.13	0.032	0.03
Plutonium	0.13	0.032	0.032
Potassium	0.75	0.180	0.180
Rhodium	0.24	0.058	0.058
Selenium	0.32	0.077	0.077
Silicon	0.71	0.17	0.17
Silver	0.23	0.057	0.057
Sodium	1.21	0.29	0.29
Tantalum	0.14	0.034	0.034
Thorium	0.13	0.03	0.03
Tin	0.21	0.054	0.05
Titanium	0.54	0.125	0.13
Tungsten	0.13	0.032	0.03
Uranium	0.12	0.028	0.028
Vanadium	0.39	0.116	0.116
Zinc	0.39	0.093	0.09
Wrought Iron	0.50	0.12	0.12

APPENDIX-XIV

RADIATION CONSTANT OF SOME COMMON MATERIALS

Product	*Radiation Constant*	
	(10^{-8} W/m² °C⁴)	*(10^{-8} Btu/(h ft² °F⁴))*
Black body	5.7	0.173
Brass, dull	0.152	0.036
Brick	5.16	0.156
Cast iron, rough oxidized	5.09	0.154
Copper, polished	0.119	0.028
Cotton	4.23	0.128
Glass	5.13	0.155
Lamp black	5.16	0.156
Oil paint	4.30	0.130
Paper	4.43	0.134
Plaster	5.16	0.156
Sand	4.20	0.127
Shavings	4.10	0.124
Silk	4.30	0.130
Silver	1.19	0.0046
Tin	0.26	0.0077
Water	3.70	0.112
Wood	4.17	0.126
Wool	4.30	0.130
Wrought iron, dull oxidized	5.16	0.156
Wrought iron, polished	1.55	0.047
Zinc, dull	0.152	0.036

σ (Stefan-Boltzmann Constant = 5.6703 10^{-8} (W/m^2K^4) = 0.1714 10^{-8} (Btu/(h ft^2 $°R^4$) ε = emissivity coefficient of the object (ε = 1 for a black body)

The radiation constant is the product between the Stefan-Boltzmann constant – σ – and the emissivity constant – ε – of the material. The emissivity coefficient – ε – indicates the radiation of heat from a **'grey body'** according the **Stefan-Boltzmann Law,** compared with the radiation of heat from a ideal **'black body'** with the emissivity coefficient ε = 1.

APPENDIX-XV

EMISSIVITY COEFFICIENT OF SOME MATERIALS

Surface Material	*Emissivity Coefficient* - ε -
Alloy 24ST Polished	0.9
Alumina, Flame sprayed	0.8
Aluminum Commercial sheet	0.09
Aluminum Foil	0.04
Aluminum Commercial Sheet	0.09
Aluminum Heavily Oxidized	0.2 – 0.31
Aluminum Highly Polished	0.039 – 0.057
Aluminum Anodized	0.77
Aluminum Rough	0.07
Antimony, polished	0.28 – 0.31
Asbestos board and paper	0.94
Asphalt	0.93
Basalt	0.72
Beryllium	0.18
Beryllium, Anodized	0.9
Bismuth, bright	0.34
Black Body Matt	1.00
Black Parson Optical	0.95
Black Silicone Paint	0.93
Black Epoxy Paint	0.89
Black Enamel Paint	0.80
Brass Dull Plate	0.22
Brass Rolled Plate Natural	0.06

Brass Polished	0.03
Brass Oxidized 600°C	0.6
Brick, red rough	0.9
Brick, fireclay	0.75
Cadmium	0.02
Carbon, not oxidized	0.81
Carbon filament	0.77
Carbon pressed filled surface	0.98
Cast Iron, newly turned	0.44
Cast Iron, turned and heated	0.60 – 0.70
Chromium polished	0.08 – 0.36
Concrete	0.85
Concrete tiles	0.63
Cotton Cloth	0.77
Copper electroplated	0.03
Copper heated and covered with thick oxide layer	0.78
Copper Polished	0.023 – 0.052
Copper Nickel Alloy, polished	0.059
Glass	0.92
Glass, pyrex	0.85 – 0.95
Gold pure and highly polished	0.018 – 0.035
Granite	0.45
Gypsum	0.85
Ice	0.97
Inconel X Oxidized	0.71
Iron polished	0.14 – 0.38
Iron, plate rusted red	0.61
Iron, dark gray surface	0.31
Iron, rough ingot	0.87 – 0.95
Lampblack paint	0.96
Lead pure unoxidized	0.057 – 0.075
Lead Oxidized	0.43

Magnesium Oxide	0.20 – 0.55
Magnesium Polished	0.07 – 0.13
Marble White	0.95
Mercury liquid	0.1
Mild Steel	0.20 – 0.32
Molybdenum polished	0.05 – 0.18
Nickel, elctroplated	0.03
Nickel, polished	0.072
Nickel, oxidized	0.59 – 0.86
Nichrome wire, bright	0.65 – 0.79
Paper offset	0.55
Plaster	0.98
Platinum, polished plate	0.054 – 0.104
Porcelain, glazed	0.92
Paint	0.96
Paper	0.93
Plaster	0.92
Plastics	0.91
Quarts glass	0.93
Roofing paper	0.91
Rubber, hard glossy plate	0.94
Rubber Nat Hard	0.91
Rubber Nat Soft	0.86
Sand	0.76
Sawdust	0.75
Silicon Carbide	0.83 – 0.96
Silver Polished	0.02 – 0.03
Steel Oxidized	0.79
Steel Polished	0.07

Stainless Steel, weathered	0.85
Stainless Steel, polished	0.075
Stainless Steel, type 301	0.54 – 0.63
Steel Galvanized Old	0.88
Steel Galvanized New	0.23
Tile	0.97
Tin unoxidized	0.04
Titanium polished	0.19
Tungsten polished	0.04
Tungsten aged filament	0.032 – 0.35
Water	0.95 – 0.963
Wood oak	0.91

APPENDIX-XVI

MOODY'S DIAGRAM FOR THE DETERMINATION OF FRICTIONAL FACTOR

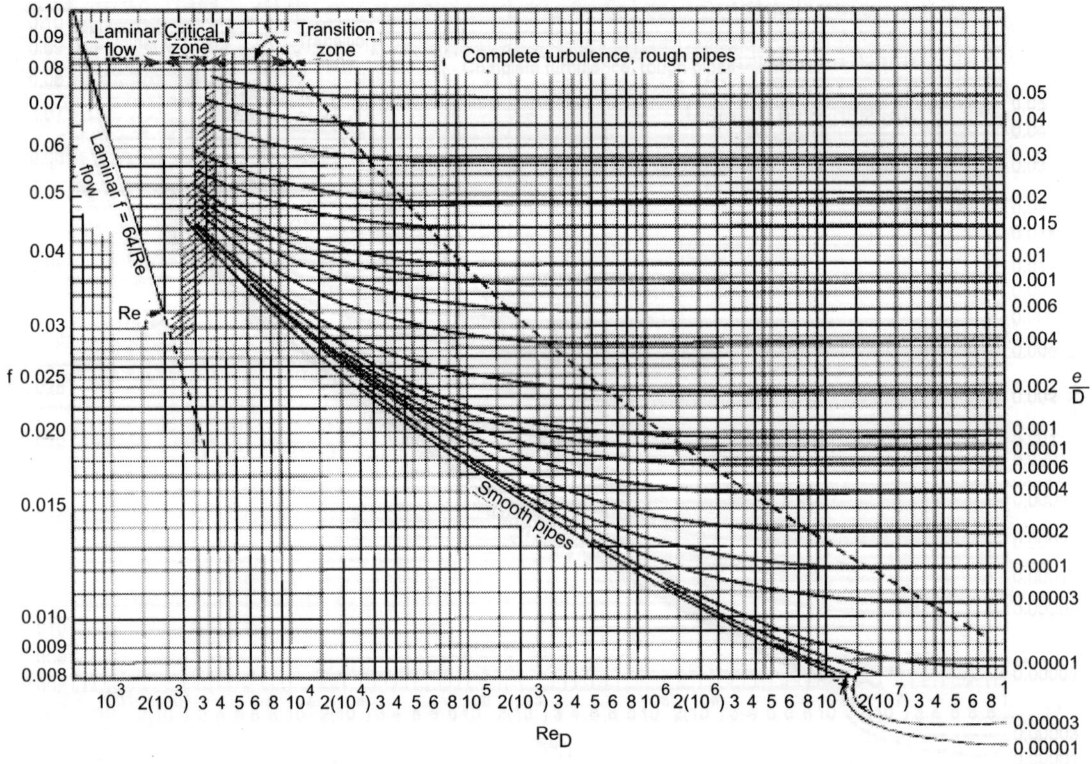

APPENDIX-XVII

TABLE FOR ABSOLUTE ROUGHNESS OF THE PIPE

Material	*Absolute Roughness Coefficient* – *k* –	
	(m) 10^{-3}	***(feet)***
Copper, Lead, Brass, Aluminum (new)	0.001 – 0.002	$3.33 - 6.7 \times 10^{-6}$
PVC and Plastic Pipes	0.0015 – 0.007	$0.5 - 2.33 \times 10^{-5}$
Stainless steel	0.015	5×10^{-5}
Steel commercial pipe	0.045 – 0.09	$1.5 - 3 \times 10^{-4}$
Stretched steel	0.015	5×10^{-5}
Weld steel	0.045	1.5×10^{-4}
Galvanized steel	0.15	5×10^{-4}
Rusted steel (corrosion)	0.15 – 4	$5 - 133 \times 10^{-4}$
New cast iron	0.25 – 0.8	$8 - 24 \times 10^{-4}$
Worn cast iron	0.8 – 1.5	$2.7 - 5 \times 10^{-3}$
Rusty cast iron	1.5 – 2.5	$5 - 8.3 \times 10^{-3}$
Sheet or asphalted cast iron	0.01 – 0.015	$3.33 - 5 \times 10^{-5}$
Smoothed cement	0.3	1×10^{-3}
Ordinary concrete	0.3 – 1	$1 - 3.33 \times 10^{-3}$
Coarse concrete	0.3 – 5	$1 - 16.7 \times 10^{-3}$
Well planed wood	0.18 – 0,9	$6 - 30 \times 10^{-4}$
Ordinary wood	5	1.67×10^{-3}

REFERENCES

1. R.H Perry, D. Green, *Perry's chemical engineer's handbook*, 6th edition, McGraw-Hill book Company, Singapore, 1984.
2. W.L. Mc- Cabe , J.T.Smith, P.Harriott, *Unit operations of chemical enginnering*, 5th edition, McGraw-Hill Book company, Singapore, 1993.
3. M.E Aulton, *Pharmaceutics*, ELBS Edition, Churchill Livingstone Ltd, Hongkong, 1990.
4. W.L. Badger, J.T Banchero, *Introduction to chemical engineering*, Mc Graw-Hill Book Company, Singapore,1955.
5. E.A. Raswlis, *Bentley's Textbook of pharmaceutics*, 8th edition, Bailiere Tindal, London, 1977.
6. S.J Carter, *Cooper and Gunn's Tutorial pharmacy*, 6th edition, Kothari Book Depot, Bombay, 1972.
7. L.Lachman, H.A. Liberman, J.L Kanig, *The theory and practice of industrial pharmacy*, 3rd edition, Lea & Febiger, Philadelphia,1985.
8. Martin, J.Swarbrick, A.Cammarata, *Physical Pharmacy*, 3rd edition, Varghese publishing house, Bombay, 1991.
9. C.V.S Subramanyam,J.Thimmasetty,Sarasija suresh, V KusumDevi, *Textbook of pharmaceutical engineering*, Vallabh Prakashan, Delhi, 2000.
10. C.V.S Subramanyam, *Textbook* of *Physical Pharmaceutics*, 2nd edition, Delhi, 2000.
11. *Indian pharmacopoeia*, 3rd, 4th and 5th editions, the Controller of Publications, Delhi, 1985, 1996 and 2005.
12. Richard E. Mark Jens Borch Charles Habeger, *Hand book of physical testing of paper*, Volume 2, 2nd edition, Marcel dekker Inc, 2002.
13. E.Shashi Menon, *Piping calculation Manual*, 1st edition, Mc Graw-Hill Book Company, Singapore, 2005.
14. www. S-ohe.com
15. www.separationprocess.com
16. www.engineeringtoolbox.com

INDEX